T5-AGA-243

COORDINATION: WATER RESOURCES AND ENVIRONMENT

PROCEEDINGS
OF SPECIAL SESSION OF ASCE'S
25TH ANNUAL CONFERENCE ON WATER RESOURCES
PLANNING AND MANAGEMENT
AND THE 1998 ANNUAL CONFERENCE ON
ENVIRONMENTAL ENGINEERING

June 1998
Chicago, Illinois

EDITED BY
William Whipple, Jr.

CO-SPONSORS OF THIS SPECIAL SESSION:
U.S. Army Corps of Engineers
U.S. Environmental Protection Agency
American Water Resources Association

1801 ALEXANDER BELL DRIVE
RESTON, VIRGINIA 20191–4400

Abstract: This proceedings, *Coordination: Water Resources and Environment,* consists of papers presented at ASCE's 25th Annual Conference on Water Resources Planning and Management and the 1998 Annual Conference on Water Resources Planning and Management and the 1998 Annual Conference on Environmental Engineering held in Chicago, Illinois, in June 1998. These papers explore the basic principles that should guide the coordination between water resources development and environmental mandates. These basic principles are: 1) maximizing tangible direct benefits from the project, as compared to the tangible costs; 2) saving lives; 3) determining environmental effects; 4) protecting the liberty of people to lead the lives they wish without undue disturbance; and 5) ascertaining human welfare effects of water resource development. While these principles are not new, what is new is the concept of considering all of them with no priority. This proceedings will assist the water resources and environmental engineering communities in determining the seriousness of the problem, the effectiveness of present national institutions, and the need for any major changes.

Library of Congress Cataloging-in-Publication Data

Conference on Water Resources Planning and Management (25th: 1998: Chicago, Ill.)
Coordination: water resources and environment: proceedings of special session of ASCE's 25th Annual Conference on Water Resources Planning and Management and the 1998 Annual Conference on Environmental Engineering, Chicago, June 1998/William Whipple, Jr., editor.
p. cm.
"Co-sponsors of this special session: U.S. Army Corps of Engineers, U.S. Environmental Protection Agency, and the American Water Resources Association."
ISBN 0-7844-0338-4
1. Water-supply–United States–Management–Congresses. 2. Environmental management–United States–Congresses. I. Whipple, William, 1909- . II. United States. Army. Corps of Engineers. III. United States. Environmental Protection Agency. IV. American Water Resources Association. V. Conference on Environmental Engineering (1998: Chicago, Ill.) VI. Title.
TD353.C63 1998 98-3518
333.91'15'0973–dc21 CIP

Library of Congress Catalog Card No: 98-3518 ISBN 0-7844-0338-4
Manufactured in the United States of America.

Contents

After the conference an addendum will be added which will consist of the conclusions of the coordination steering committee as to the need for further procedures to obtain coordination between water resources and environmental objectives.

FIRST PLENARY SESSION

Moderator, William Whipple, Jr., FASCE

"Introduction to Coordination: Views Which Should Govern"
William Whipple, Jr.
Greeley-Polhemus Group

"States Water Resources Planning and the Settings"
Kyle Schilling
Director, Institute for Water Resources, Corps of Engineers

"EPA Views on Coordination"*
Robert Perciasepe
Assistant Administrator for Water, Environmental Protection Agency

*Text not available at time of printing.

Guiding Principles for Coordination

William Whipple, Jr., F ASCE[1]

Abstract

In introducing the subject of obtaining coordination between water resources development and environmental mandates, it is necessary to consider what principles should guide us. We cannot simply rely on existing legislation, because a main purpose of the conference is to consider possibilities of new legislation to achieve that coordination. Economic, environmental and other intangible principles are equally applicable.

I am pleased and honored to be leadoff speaker in an important part of this conference, dealing with the necessity for coordination between water resources development and environmental mandates. This is an important matter because of the changed conditions in which we now live. The era of wholesale building of large dams is now past, and we are in an era of problem-solving, where, although some dams may still be built, many different kinds of actions must be considered. These may include relicensing of dams, using computers to schedule flows more efficiently, reallocating water to different uses, conserving or reclaiming water, regulating the use of water, and changing structural features to make old projects work better. The interests of both the divisions responsible

---

[1]Principal, Greeley Polhemus Group, 395 Mercer Road, Princeton, NJ 08540.

for the conference, the Water Resources Planning and Management Division and the Environmental Division, are involved in this coordination problem. We have cosponsorship of the conference by the U.S. Corps of Engineers, the EPA and the American Water Resources Association.

It may be interesting to mention briefly how the ASCE happened to hold a conference on coordination of environmental interests and water resources planning. Based on my experience, I had written a small book and a paper concluding that such coordination was necessary (Whipple, May/June, 1996; Whipple, 1996). I then suddenly realized that more action by more people was required, and began calling influential old friends and associates round the country. They are today the coordination steering committee. With their advice and support, the ASCE became interested; and this conference was arranged. We hope that this conference will result in a general consensus in the profession as to whether or not we need new congressional action and institutional arrangements to improve the situation, and if we do, what arrangements.

At this time, before we get into specific problems and situations, I think it would be useful to consider the general principles and objectives which should guide us as we consider coordination between water resources planning and environmental mandates. Those of us who have been in government service are accustomed to considering federal legislation and implementing regulations as the basis for all decision-making. But for the issues we are discussing this week, this is not sufficient because different federal agencies are involved and because we are considering the possibility and the desirability of federal legislative changes. These principles and objectives are not new; they will all be familiar to you. What is new is the concept of considering all of them, with no priority. These principles and objectives which should guide us are outlined in somewhat greater detail in a book which has just appeared (Whipple, 1998).

The first objective[2] is to maximize the tangible di-

---

[2]The order in which the principles are given does not indicate priority.

rect benefits from the project, as compared to the tangible costs of building, maintaining and administering it. Since most of the costs are incurred at the time of construction of the project, while the benefits occur over a period of many years after the project is finished, a discount rate (usually the rate of return of long-term federal bonds) is used to compare dollar value of costs at time X to value of benefits at time Y.

OBJECTIVES OF WATER RESOURCE PLANNING
(order given does not indicate priority)

1. Maximize tangible benefits less costs
2. Saving of lives
3. Environmental effects (non-human)
4. Liberty to live without disruption
5. Welfare effects on human life, including traditional, historical, religious, and archeological aspects

The second important factor is the saving of lives. This point arises mostly in flood control projects, which may prevent loss of life. Sometimes attempts are made to put a money value on human life. This is unacceptable in principle, but rough approximate evaluations are sometimes made.

The third group of objectives and problems is the broad category of environmental effects. These effects may be benefits but often are projected adverse effects of the proposed (or actual) project. These include effects on fish and wildlife generally, endangered species of plants, fish, birds and animals, and also the value of the area for public recreation. The evaluation of environmental effects is an enormously complex matter. However diverse the various explanations and justifications for the environmental beliefs, those beliefs are strong and (mostly) sincere. Moreover, they have a lot of political support. Of the nations on earth, the United States is one of the most advanced in environmental protection, and there is no sign that this position will change. Our natural environment can and must be protected.

The fourth important factor is the liberty of people to lead the lives they wish without undue disturbance. In many parts of the country families have their lives disrupted by prolonged drought or by floods, which destroy their way of life and may force them to leave and live elsewhere. More rarely, a similar disruption may occur through erosion. In many cases, water resources developments can provide safety against such eventualities. There are many more less disastrous but still serious events which flood control and assured supplies of water can guard against. Prudent families are usually prepared to face unexpected losses of moderate amount, but to protect against major losses such as destruction of an automobile, or death of a wage earner, prudent families buy insurance, paying a great deal more, often twice as much,as the financial value of the risk avoided (otherwise insurance companies would lose money). In times of flood and sometimes in water supply and irrigation situations, projects may provide safety against unpredictable major losses. As is the case with insurance, this protection provides benefits greater than the money value of the anticipated loss.

The fifth category, not usually considered in a benefit-cost analysis is simply human welfare effects of water resource development. This includes full employment, public health, and education of children, all of which can be affected by projects.

Some of the effects which must be considered, where they exist, result from change in the condition of groundwater. Groundwater is sometimes unduly depleted by project withdrawals, or benefited by sustained low flows. Occasionally, the effect upon groundwater is the primary purpose of a project.

Sometimes, projects have a significant effect upon water quality. This may be favorable, such as allowing development of a trout fishery just below a dam, or the removal of sedimentation. On the other hand, variability of flows downstream of power dams may be harmful to fish.

If the outlet to a reservoir were designed to release water only from the lower depths of the reservoir, water quality could be adversely affected.

Both groundwater and water quality effects may have tangible economic aspects or may be manifested as environmental intangibles.

An effect which is often important, especially in the West, is the effect upon the interests, beliefs, and traditions of Native Americans. In building the Columbia River dams, the tribes originally opposed the program because of the interference with traditional fishing of salmon at the falls, which were to be inundated. There was also the question of native burial grounds, which were to be flooded by the reservoirs. These matters were given very careful consideration to minimize the problems, but some problems still remain.

Sometimes historical and archeological interests may be involved, or the taking of a church or cemetery.

All of these matters are relevant to evaluation of a project, to the extent that they apply to the particular case; and the design of the project must be adapted to them. Only the first category, the tangible benefits and costs, can be evaluated reliably in dollars and cents. Other categories can be used in cost comparisons by determining which is the least-cost way of obtaining an important intangible objective. The consideration of all these intangibles is essential to the solution of today's problems.

The valuation of environmental objectives covers a number of subjects. Early aspects of the environmental movement centered on creating great national and state parks and forests, and protecting forests generally. The next great movement was to treat major sources of pollution under the Clean Water Act. More recent views include the concept of biodiversity, the value of protecting wetlands and preserving endangered species. Environmental changes both upstream and downstream from dams, including environmental advantages conferred by dams and flood plain management, must all be considered.

These various environmental principles and objectives are important national objectives, but the question remains: Are they absolute or relative values? My view is that they are all relative values. Our Constitution aims to provide to each man life, liberty and the pursuit of happiness, but even the constitutional aims are not absolutes. The right of a criminal to liberty and the pursuit of happiness is severely limited in order to protect the rights of others. Similarly, the objective of preserving some endangered species or reducing pollution to the lowest possible levels should be considered relative to other worthwhile national objectives. There should be no easy automatic answers to such questions. To automatically favor environmental interests above all others would be as wrong as the old practice of ignoring them.

In the new era, these difficult problems must now be faced. That is what this conference is all about. We have to decide (a) how serious is the problem? (b) How far can the matter be handled with our present national institutions? (c) If a major change is desirable, what kind of changes should be made?

This is a great challenge. The next few days will show whether or not we are worthy of it.

References

Whipple, W. "Integration of Water Resources Planning and Environmental Regulation." Journal Water Resources Planning and Management. ASCE May/June 1996.

Whipple, W. Comprehensive Water Planning and Regulation: New Approaches for Workable Solutions. Government Institutes Inc. Rockville, Md. 1996.

Whipple, W. "Water Resources: The New Era for Coordination." ASCE, 1998.

States Water Resources Planning and the Settings

Kyle E. Schilling, P.E., DEE[1]

Abstract

The past quarter century of water resources planning practice in the United States has demonstrated a shift from a well codified federal system to a new, more inclusive approach emphasizing quality and the environment. New planning approaches and models used in shared vision and incremental cost planning (STELLA, ECO-EASY, and IWR-PLAN) will be described along with case applications which can help to resolve conflicts with entrench vested interests.

General

The 25th anniversary of Public Law 92-500, the Water Pollution Control Act (CWA) occurred in 1997. The CWA was part of a broader shift to reliance on regulation by the Federal government in this same time period. According to the Advisory Commission on Intergovernmental Relations, 53% of the 439 federal preemption statues, since the Constitution was ratified, were enacted in the 21 years from 1969 through 1991. That trend continues. The traditional federal benefit-cost analysis used for traditional water resources planning was perceived as undervaluing water quality and environmental issues. Yet, by most standards, the most obviously beneficial and least controversial projects had been built by the late 1960's explaining the diminishing rate of dam construction. The federally-funded Water Resources Council and many river basin commissions set up to assist in these planning processes are now also gone. The last quarter century has been dominated by a transition from traditional water quantity development approaches to water quality issues. Observers note that costs and effort are now increasing relatively for each new increment of gain. Debate over each new project is now replete with references to regulatory reform, unfunded mandates, benefit cost analyses, etc. This situation is analogous to the history of the nation's water resources development program where broad support for action existed when benefits were obvious, widespread and when federal financial shares were larger. Federal policies came under increasing debate when benefits were relatively smaller, more site specific and judged to be of little federal financial interest.

----

[1]Director, Institute for Water Resources, U.S. Army Corps of Engineers. The views expressed by the author are entirely his own based on over 30 years of water resources planning and management experience in a variety of Federal and state positions and do not reflect the view of the U.S. Army Corps of Engineers nor the United States Government.

Many emerging water resources needs may be characterized as reengineering hydrologic variability into the water and land environment. Within agencies such as the Corps of Engineers new environmental authorities have also been added to restore fish and wildlife habitats at existing Corps projects and numerous such efforts are now underway.

At this point in time, many major water resources works are now 40-50 years old and have been well integrated into our economic, legal and social fabric. When built, they were planned to meet distant future needs which they have done well, but they have also proven remarkably resilient in managing to meet changing needs. That capacity for adaptation still exists and will continue to serve the U.S. well, with proper attention to improved operations, maintenance and rehabilitation. Not many major new works are expected; smaller, new works for more localized quantity and quality issues or more efficient system operations of existing projects are now expected to be the predominant activity with emphasis on quality and the environment.

We developed and fine tuned federal multi-objective, multipurpose planning procedures under the Water Resources Council's Principles and Standards and later Principles and Guidelines for Water Resources Planning based on benefit-cost analysis for water quantity planning over the last 25 years. These include public involvement, economic, social and environmental analysis. These procedures provided for explicit planning objectives, Environmental Quality (EQ), Regional Development (RD) and Other Social Effects (OSE) and multiple purposes such as water supply, navigation and flood control. Yet the federally dominated, multi-objective, benefit cost approach for water quantity has been overtaken in national significance by a single purpose (water quality), single objective (EQ) regulatory approach, relying on local action to meet Federal standards. This philosophical shift in approaches appears to reflect both the maturity of the nation's water quantity development program and that the traditional planning process was seen as undervaluing the environment. The greater effort expended to develop good procedures also reflected the decision makers and public's desire to make sure we were doing the right things since the most needed development had taken place, therefore creating tougher tests for additional investment. However, significant flood, drought and water allocation management problems remain in the U.S. despite the well-demonstrated management flexibility provided by the existing system. Addressing those problems will require improved management with increased attention to overall environmental and environmental restoration interfaces.

As an example of change activities, the Institute for Water Resources is working hard to develop techniques and new tools for the Corps of Engineers to focus on environmental sustainability and decision making.

This increased emphasis on the environment, however, brings with it a need for improved techniques for evaluating the comparing environmental projects and programs.

Until recently, there has been a lack of accepted methods for assessing the effectiveness (does the project achieve its objective?) and efficiency (is it achieved in the least cost manner?) of investments in the protection or restoration of environmental resources. To addresses these issues, the Corps initiated the Evaluation of Environmental Investments Research Program (EEIRP).

The Corps water resources program has changed significantly over the past two decades. These changes reflect changing national preferences and desires. Alteration of watersheds for such purposes as flood control and navigation is no longer considered a sure path to economic development. There is more concern today for the protection and restoration of the natural services of heavily altered watersheds, many of which were related to previous Corps water resource development projects.

The EEIRP

The overall objective of the EEIRP was to provide an evaluation framework, techniques, and procedures to assist planners, managers and regulators in addressing both the site and portfolio issues: i.e., whether the recommended action is the most effective and efficient alternative for a particular location, and how to allocate limited resources among competing recommended actions. One goal of the program is the development of a series of environmental evaluation procedures manuals ("how to") manuals) addressing various steps in the planning, evaluating and prioritizing processes. To accomplish these objectives, the research program has been divided into ten more specific areas called work units. The objectives of each of these studies are listed below:

Evaluation of Environmental Investments Procedures Manual Interim: Cost Effectiveness and Incremental Cost Analyses, IWR Report 95-R-1, May 1995.

Review of Monetary and Nonmonetary Valuation of Environmental Investments, IWR Report 95-R-2, February 1995.

Prototype Information Tree for Environmental Restoration Plan Formulation and Cost Estimation, IWR Report 95-R-3, March 1995.

Compilation and Review of Completed Restoration and Mitigation Studies in Developing an Evaluation Framework for Environmental Resources, Volumes I and II, IWR Reports 95-R-4 and 95-R-5, April 1995.

Trade-off Analysis for environmental Projects: An Annotated Bibliography, IWR Report 95-R-8, August 1995.

Resource Significance: A New Perspective for Environmental Project Planning, IWR Report 95-R-10, June 1995.

Trends and Patterns in Cultural Resource Significance: A Historical Perspective and Annotated Bibliography, IWR Report 96-EL-1, January 1996.

National Review of Non-Corps Environmental Restoration Projects, IWR Report 95-R-12, December 1995.

Linkages between Environmental Outputs and Human Services, IWR Report 96-R-4, February 1996.

An Introduction to Risk and Uncertainty in the Evaluation of Environmental Investments, IWR Report 96-R-8, March 1996.

National Drought Study and Advanced Modeling Techniques

Another adaptation process in which the Institute is involved is the National Drought Study. Drought planning 25 years ago was almost always aimed at new supply development; it's now focusing on environmentally friendly demand management. In this study, plans are designed to take advantages of advances in the component skills of drought management, hydrology, water use forecasting, public involvement, multi objective planning and evaluation, local governmental and decision support modeling to capture and implement a way of behaving. This contrasts with traditional federal planning which might place more emphasis on structures or reservoir rule curves. The Drought Study developed and tested an integrated approach to water management called the DPS Method. The approach is a suitable foundation for watershed, river basin, adaptive or ecosystem management. It is based on the best traditional planning and evaluation principles, modified to reflect the increasingly non-federal, non-structural nature of water management.

The most visible innovation from the Drought Study is the use of a new generation of user friendly software to create shared vision models. Shared vision models bridge the gap between specialized computer models and the way people conceptualize problems and make decisions. The name, shared vision models, captures their most important advantage. Experts and stakeholders can build these models together, including elements that interest each group to build a consensus view of how the water system works as a whole and how it affects stakeholders and the environment.

IWR has applied the techniques developed during the National Drought Study in the Alabama-Coosa-Tallapossa (ACT) and Apalachicola-Chattahooche-Flint (ACF) Comprehensive River Basins Study. The study was conducted by Alabama, Florida, George and Mobile Corps District (study partners).

The purposes of the study were to:

• determine the capabilities of the water resources of the ACT and ACF basins;

• describe various water resource demands in Alabama, Florida and Georgia within these basins, and

• to evaluate alternatives for the management and utilization of the water resources to benefit all user groups within the basins.

Early in 1994, study partners asked IWR to conduct the basin wide management element of the study using the 7-step "shared vision" planning and evaluation method developed during the National Drought Study. The basin wide management study draft has been completed. IWR has worked with the states to develop a statement of problems, planning objectives, performance measures, constraints and decision criteria; describe a reference condition; and identify and evaluate alternatives through "shared vision models." The agreement by the parties in the study to form a new river basin organization based on these "shared vision" concepts is a milestone.

The ECO-EASY software for conducting cost effectiveness and incremental cost analyses in environmental planning studies was developed as part of the EEIRP program. The procedures are useful for formulating alternatives plans, identifying which of those plans are cost effective and conducting incremental cost analysis. The results of the analyses help planners and decision makers address the question "how much environmental benefit is worth its cost?" IWR and the Corps Waterways Experiment Station have incorporated these procedures into a software program called ECO-EASY.

The application of this cost effectiveness and incremental cost methodology is becoming widespread across the Corps ecosystem restoration program. Recent advancements in the form of instructional manuals and the ECO-EASY software have improved the ease and speed of the analyses for field practitioners. A recent Corps field application is documented as a case study involving fishery habitat improvement at Bussey Lake, Illinois.

Recent Corps experiences indicate that the analyses are applicable to both environmental restoration and mitigation planning; that they are useful for a wide range of sizes of problems and projects; and that they can be used to scope solutions even at the earliest stages of planning. In addition, although the analyses have thus far focused on fish and wildlife habitat and ecosystem related studies, they should be equally useful in addressing other environmental problems such as water and air pollution and hazardous waste. Other agencies have indicated the potential applicability of the procedures to a wide range of problem solving scenarios including the ordering of Superfund cleanup sites and transportation alternatives analysis.

ECO-EASY conducts three processing functions: formulation of combinations, cost effectiveness analysis of combinations, and incremental cost analysis of cost effectiveness combinations.

The Institute for Water Resources in partnership with the Natural Resources and Soil Conservation Service have now developed IWR-PLAN Decision Support Software to assist with formulating and comparing alternative plans. IWR-PLAN can assist with plan formulation by combining solutions to planning problems and calculating the additive effects of each combination, or "plan." IWR-PLAN can assist with plan comparison by conducting cost effectiveness and incremental cost analyses, identifying the plans which are the best financial investments and displaying the effects of each on a range of decision variables.

IWR-PLAN builds upon the basic plan formulation and comparison framework of the DOS program ECO-EASY, developed within the Corps Evaluation of Environmental Investments Research Program. The IWR-PLAN system transforms ECO-EASY to a Windows 95 or Windows NT operating environment while adding new functions. Development of IWR-PLAN has been carried out within the Corps Decision Support Technologies Research Program, Conducted at IWR.

IWR-PLAN Beta Version 1.5 was completed in 1997 and is available via the IWR-PLAN download site. The IWR-PLAN home page provides access to the download site as well as a checklist of operating instructions, hints and suggestions, and a news board for announcing program developments, such as upgrades (http://www.wrc.ndc.usace.arm.mil/iwr/iwrplan/iwrplan.htm). The 1998 beta-testing and feedback period includes several scheduled upgrades, providing additional refinements and new functions.

The system will formulate alternative combinations of solutions and compare their effects on up to ten user specified decision parameters. Derived parameters can be defined that are weighted combinations of other decision variables. Constraints can be set, specifying minimum and maximum acceptable values for each decision variable. IWR-PLAN's sensitivity module allows examination of the implications of uncertainty in decision variable estimates. Plans of interest can be identified and displayed throughout the analyses regardless of their cost effectiveness.

Multiple scenarios can be examined from a single set of input data. Scenarios may differ as to what decision variables are included in the cost effectiveness analyses, what solutions are included, what sensitivity values are used, what groups of constraints are applied, and what plans of interest are included. The results form different scenarios can be compared through IWR-PLAN's multiple scenario comparison module.

A variety of graphing and reporting options are available. All graphs and data in an IWR-PLAN file are directly exportable to a range of other programs to assist with reporting. IWR-PLAN also comes with an on-screen help system which provides instruction for all forms and functions in the program.

The foregoing simple, yet powerful tools in conjunction with IWR's recently published Planning Manual and Planning Primer appear to be serving the needs of a large pool of pent up demand for a return to basic planning, supplemented by new tools. IWR staff has been requested to run, on a reimbursable basis, a large number of orientation and training workshops for the Corps, other federal agencies and levels of government.

Conclusions

Pressure is going to reintegrate water quantity and quality planning, with greater emphasis on both the environment and overall performance including cost. The nature of the planning tools and decision processes to support these changes is undergoing rapid and fundamental change to become more interactive and user accessible; quickly show impacts and trade-offs between alternatives; and provide a larger decision making framework, integrating the environment.

The engineering community involved in water resources planning and management has an opportunity to help reinvigorate the practice of good planning if it can itself adopt to the new decision making environment.

References

Advisory Commission on Interngovernmental Relations (1992), *Intergovernmental Decisionmaking for Environmental Protection and Public Works.*

Appelbaum, Stuart J. (1996) *THE C&SF Project Comprehensive Review Study: Interagency Planning Team Integration*, Watershed '96.

Living within Constraints: An Emerging vision for High Performance Public Works, June 1995.

Schilling, Kyle E. (1995), *A Perspective on Untied States Water Resources and Institutions Related to Environment*, paper.

Stakhiv, Eugene Z. (1989), *Vision 21,* U.S. Army Corps of Engineers, Institute for Water Resources.

Whipple, William Jr. (1996) *Integration of Water Resources Planning and Environmental Regulation*, Journal of Water Resources Planning and Management, ASCE, May/June 1996 issue.

Second Plenary Session

Moderator, William Whipple, Jr., F ASCE

"History and Background of Water Resources Planning"
Neil Grigg, Chairman, Department of Civil Engineering, University of Colorado

"The States' Viewpoint"
Donald R. Vonnahme, Director, Office of Water Resources, Illinois Department of Natural Resources

Water Resources Development and Environmental Protection: Background and Principles

Neil S. Grigg, F. ASCE[1]

Abstract

The paper describes the evolution of the institutional setting which frames water resources management in the United States today. Most of the issues to be discussed at the conference occurred from about 1900 until the present when the number of stakeholders increased greatly, and attempts to formulate workable processes for planning and coordination have met complexity, conflict, and resistance. The difficult part is learning how to balance values in a complex society, with water decisions being especially difficult because of the high levels of interdependency. Today, decisions are worked out in participatory democracy, and often the public ends up paying millions of dollars for fees to lawyers and consultants, as well as staffs of regulatory and resource agencies. Finding a better approach is the main goal of this conference.

Introduction

The purpose of the paper is to describe the institutional setting which frames water resources management in the United States today. I will trace the historical development of water development and regulation, cite a few case studies that illustrate current conflicts, and pose questions that must be answered if the nation is to find a workable paradigm to improve coordination between water development and environmental regulation. The description in the paper will of necessity be brief, but further detail is available in publications such as (Grigg, 1996, Holmes, 1972, Viessman and Welty, 1985). In another paper presented at the conference, I will outline options for improving the coordination process.

---

[1] Professor and Head, Department of Civil Engineering, Colorado State University, Fort Collins, CO 80523.

Historical evolution

In the American Colonial period, water development was mostly limited to small projects. In the antebellum years of the 19th Century, water construction remained limited, although a few projects were initiated. After the Civil War, water development in the West began, mainly with irrigation projects, and the nation also started to realize how important water is for industrial and urban development. In the 20th Century, water development started to take off. From 1900 to World War II, a number of reservoir projects were completed, both public and private. After World War II, construction of reservoirs continued until the pace slowed with the current regulatory era.

Most of the issues we will discuss at this conference occurred from about 1900 until the present. Actually, you might say that 1900 to 1970 was the development era and 1970 to the present has been the regulatory era.

In the development era, the term "water resources planning" mostly meant planning for facilities that met economic goals such as hydropower and/or irrigation. Goals were usually focused on single purposes such as irrigation water, hydropower, navigation, or flood control, and the emphasis was on economic development, not environmental needs. Many Bureau of Reclamation and Corps of Engineer projects were planned and built in this era.

Early in the century, as the nation developed, it became clear that water decisions involve many industries, geographical areas, and public interest viewpoints, and the possibilities of multiple purpose development became apparent. The Flood Control Act of 1917 called for "a comprehensive study of the watershed" including the study of power possibilities, although the Corps of Engineers, the nation's main flood control agency, was resistant to the concept of multiple-purpose planning (Holmes, 1972). Thus, coordination was mainly present in agency programs to build water projects.

The Depression, and its emphasis on public works to prime the economic pumps, gave increased attention to water development. The National Industrial Recovery Act of 1933 called for a "comprehensive" program of public works to consider the full spectrum of water resources uses, which included, according to Holmes (1972): control, utilization and purification of waters, prevention of soil and coastal erosion, development of water power, transmission of electric energy, river and harbor improvements, flood control..."

Disputes grew after World War II, and a Senate Select Committee on Water Resources was appointed. The Water Resources Planning Act, first passed in 1962, served to institutionalize the term "comprehensive planning.". The Act stated that it is the "...policy of the Congress to encourage the conservation, development, and utilization of water and related land resources of the United States on a comprehensive and coordinated basis by the Federal Government, States, localities, and private enterprise..." The Act provided support for state planning programs, for the establishment of a National Water Resources Council, and for river basin (Level B) studies.

The period 1965-1980 was an active one in water resources planning. You might say that in this period the New Deal concepts of activist government involvement in water resources were tested and shown to fail. By 1981 the concepts of the Water Resources Planning Act were pretty much dead. The Carter "hit list", which dealt with both reform in government and with environmental issues, had been part of the reason for the demise. While there was considerable support for government reform and environmental preservation, the Carter Administration was thrown out due to problems such as high inflation. Today, the climate for the planning and decision process is more challenging than ever. Recent emphasis has been on tax cutting, regulatory reform, privatization, and other non-governmental initiatives.

Along the way, as the number of stakeholders increased, and as objectives multiplied, attempts were made to formulate workable processes for planning and coordination, but they met complexity, conflict, and resistance. In spite of many attempts, no consensus process for comprehensive, coordinated, joint planning exists.

Today, rather than following a rationalized "comprehensive planning process," proposals must pass a series of feasibility hurdles including technical, financial, legal, environmental, political, and judicial review. Environmental interest groups use national legislation and the courts to pursue their goals, states sometimes provide forums for coordination, national and state legislation set policy for decision-making, and interstate issues often involve complex negotiations. New emphasis on fish and wildlife and on species protection continues to render coordination more difficult. The process is worked out in the messy arena of participatory democracy, and no single person or agency has complete control of it, even for a single problem or issue.

Incidentally, the foundation for the formation of ASCE's Water Resources Planning and Management Division was laid about the beginning of the regulatory era. As the complexities of water resources planning increased, ASCE recognized that more attention to planning and policy was needed, and in 1973 the Division was formed(Committee on Water Resources Planning, 1962). The Division celebrates its 25th anniversary at this convention in Seattle in 1998.

Examples

Today, some new reservoirs are needed, but increasingly, coordination is needed for scenarios other than new project construction. A few examples of such scenarios may shed light on conflicts between development and regulation where a process is needed.

First, consider a generalized hypothetical watershed with multiple players and purposes of water management. A diagram of such a watershed formed the frontispiece nearly fifty years ago for the report of the President's Water Resources Policy Commission (1950). Water management in this watershed requires stakeholders to work together to solve shared problems. Reservoir releases should be coordinated by a joint decision body. New facilities or changes in use of water should be coordinated by a process that balances all views. Regulatory authorities should not just "command and control," but should coordinate flexible rules on health, environment, reliability, and cost of service. Good management practices should be used throughout, for example, water pricing should allocate water and avoid waste. Information technologies should provide shared information for management. This idealized situation would require an effective coordination mechanism at the watershed level.

Next, consider some actual examples of complex, difficult water management scenarios in the United States today.

California's Bay-Delta – a complex, areawide problem. California's "Bay-Delta" illustrates a very complex regional water management scenario, all in one state. According to Brickson and Sudman (1990), the issues are: balancing urban, agricultural, and environmental water uses; water quality and salinity standards; water for striped bass and salmon; water quality for 19 million Californians; flood protection and levees; and managing harmful flows such as from pumping actions." There here are so many issues that a special coordinating mechanism, CALFED, has been formed. CALFED is "the joint state-federal planning organization created in June 1994 to provide more coordinated action in the Bay-Delta." (McClurg, 1996). At this point, no one knows if CALFED will succeed.

Two Forks - a western water supply reservoir. In the Denver region, a large water supply project that had been planned for decades was vetoed by the Environmental Protection Agency, throwing the water supply planning process into turmoil in spite of an environmental study that cost over $40 million. The conflict began when Denver, joined by a group of suburban water agencies, initiated planning for a joint water supply project. Initial integration was attempted through negotiation, a Governor's water roundtable, and personal political work. The joint project seemed a triumph for regional cooperation and negotiation, but in 1992 environmentalists, who had participated in the planning process, convinced EPA to veto it. In the aftermath of Two Forks, Denver withdrew from its role as the leader of regional water efforts and will only take care of its own needs in the future. Denver refused to join a suit to overturn the Two Forks veto (Obmascik, 1991). Coordination did not work at all here; regulation was the hammer.

Platte River – a river basin with endangered species issues. On the Platte River, there has been a long term effort to relicense hydropower under authority of the Federal Energy Regulatory Commission (FERC). Endangered species issues have resulted in studies, negotiations, court actions and a recovery plan. In 1994 and 1995 a federal-state negotiating group worked to seek joint plans and actions, and in 1997, a three-state accord was signed by Colorado, Nebraska, and Wyoming (DeSena, 1997). This required federal intervention to coordinate negotiations, and the Secretary of Interior also signed the agreement. Paying for the cooperative agreement will involve $2.5 million in federal funds and $1.8 million in state funds over the next three years. Whether this agreement will work is still unknown.

Albemarle-Pamlico Sounds – an estuary water quality issue. Albemarle and Pamlico Sounds, located in northeast North Carolina, have serious environmental and water quality problems. I was involved with others in 1979 on a plan to reduce pollutant loads, but a stumbling block was lack of agreement among the states, local governments and industries about the solutions needed. More recently, the region has been studied via the Albemarle-Pamlico Estuarine Study, and environmental conflicts continue (1988). One highly visible issue is fish disease that involves microorganisms that some believe to be toxic. If there is any coordination, it is by state regulatory authorities. Could any other approach work?

Virginia Beach Water Supply – an eastern interbasin transfer issue. Virginia Beach has been seeking a secure water supply via an interbasin, interstate water transfer for nearly 20 years (Grigg, 1996). Efforts to coordinate failed. Virginia Beach needed the cooperation of North Carolina to gain permission for a planned pipeline. The two states had a bi-state committee to discuss water issues, and progress was apparent, but politics interfered, and North Carolina announced opposition to the project. Many individual disputes have occurred, and they have been referred to the regulatory and judicial arenas, not coordination forums.

These examples illustrate the complexity and interdependency that defy efforts at coordination, without regulatory authority. In the Bay-Delta, without the federal-state forum, it is hard to see how progress could be made outside of the dispute resolution process. Two Forks involved coordination that included a governor's roundtable and an EIS, but $40 million was still wasted on a failed effort. The Platte River shows attempts at coordination, and while the states have agreed on parts of the solution, this may not produce an ultimate solution, in spite of a three year negotiation process. Albemarle-Pamlico and Virginia Beach lack any integrating mechanism other than the regulatory process and adversarial actions.

Conclusions

So, the issues needing coordination reach well beyond the "project" format or earlier years, and the difficult part is learning how to balance values in a complex society, with water decisions being especially difficult because of the high levels of interdependency. The nation built many projects in the development era, but has not learned to handle the complexities of the regulatory era where the number of stakeholders has increased and no consensus process for comprehensive, coordinated, joint planning has evolved.

Today, proposals must pass a series of tests involving technology, finance, law, environment, and politics. Powerful interest groups withhold their ultimate cooperation from coordination forums, just in case things do not work out to their liking. Decisions are worked out in participatory democracy, and often the public ends up paying millions of dollars for fees to lawyers and consultants, as well as staffs of regulatory and resource agencies. Something better is needed. What should it be? Finding answers to the puzzle is a main goal of this conference.

References

Albemarle-Pamlico Estuarine Study, Albemarle-Pamlico Advocate, Vol 1, No 1, Washington, N.C., July 1988.

Brickson, Betty and Ruth Schmidt Sudman, A Briefing on California Water Issues, Western Water, September/October 1990.

Committee on Water Resources Planning, Basic Considerations in Water Resources Planning, Journal of the Hydraulics Division, American Society of Civil Engineers, HY 5, September 1962.

DeSena, Mary, Governors Sign Three-State Platte River Agreement, US Water News, Vol 14, No. 8, p. 1, August 1997.

Grigg, Neil S., Water Resources Management: Principles, Regulations, and Cases, McGraw-Hill, New York, 1996.

Holmes, Beatrice Hort, A History of Federal Water Resources Programs, 1800-1960, US Department of Agriculture, Economic Research Service, Washington DC, June 1972.

Long's Peak Working Group, America's Waters: A New Era of Sustainability, Natural Resources Law Center, University of Colorado, Boulder, December 1992.

McClurg, Sue, Delta Developments, Western Water, January/February 1996a, Water Education Foundation, Sacramento.

President's Water Resources Policy Commission, A Water Policy for the American People, Washington, 1950.

Viessman, Warren, Jr, and Claire Welty, Water Management: Technology and Institutions, Harper & Row, Inc., New York, 1985.

The Emerging State Role in Planning and Management

Donald R. Vonnahme[1]

Abstract

Water resources planning and management in the United States is in the midst of a profound transition with equally profound implications for state governments. This transition has three primary characteristics. First, it reflects a movement from a top-down, command and control, federal government dominated approach to a bottom-up, partnership-based, inclusive approach. Second, it acknowledges the need for multi-objective planning; for meeting environmental and socio-economic objectives as well as engineering objectives. And third, the transition is characterized by fiscal constraints and the attendant need for creative funding of water resources projects. Singly and collectively, these characteristics represent a new era in water resources planning and management that demands an unprecedented level of state leadership and multi-agency partnership. As a consequence, the current and potential role of interstate river basin organizations is enjoying renewed emphasis.

Introduction

This paper will explore the ongoing evolution of U.S. water resources planning and management in the context of federal/state relations; state roles and responsibilities; and the growing relevance of interstate river basin organizations. The trends and characteristics of what might be termed a "new era" in water resources planning and management will be presented, along with the attendant implications for state governments. The "institutional infrastructure" needed to facilitate the transition to this new era—principally interstate river basin organizations arrangements—will be discussed. Examples of several such arrangements will be presented, with an emphasis on the Great Lakes Basin. Concluding remarks will identify challenges

---

[1]Director, Office of Water Resources, Illinois Dept. of Natural Resources and Chair, Great Lakes Commission, 524 S. Second St., Springfield, IL 62701

and opportunities for state government, and offer several recommendations that might serve as a blueprint for the future.

The Emergence of a New Era in Water Resources Planning and Management: Trends and Characteristics

Water resources planning and management in the United States has a rich and varied history, and might be described as a grand, continuing experiment in intergovernmental relationships and institutional design. Its evolution can be characterized by a series of eras, each with its own distinct planning and management emphasis, and attendant state role. (Donahue, 1996) The current era has its roots in the early-mid 1980s and continues to unfold today. The underpinnings of this era are found at both the federal and state levels. With regard to the former, the "new federalism" philosophy of the Reagan Administration viewed water resources issues primarily as state or interstate concerns, thereby encouraging solutions at that level. A "kinder and gentler" federal government emerged in the 1990s, tempering its regulatory emphasis with voluntary compliance and partnership approaches. And, today, the downsizing and "re-invention" of the federal government continues, prompted by efficiency concerns and budgetary realities. At the state level, the last decade or so has seen a steadily rising ethic of self determination in all aspects of governance, accompanied by a stronger sense of resource stewardship and an attendant need for interstate collaboration on regional issues and management needs.

In addition to shifts in political and governance philosophy, the present era is also characterized by a fundamental shift in planning and management philosophy. Multi-jurisdictional, watershed-based planning and management is enjoying new-found popularity. The "ecosystem approach" continues to move from concept to application, entering the mainstream and markedly broadening the diversity of participants, viewpoints and objectives associated with planning and management efforts. More recently, the related notion of "sustainable development" has been embraced (at least conceptually) by all levels of government, and has fostered an unprecedented level of public/private partnership. (Great Lakes Commission, 1990)

The characteristics and trends of the present era might best be summarized by the following transition in planning and management philosophies:

"top-down" mandates	→	"bottom-up" initiatives
command and control regulatory emphasis	→	partnership-oriented, voluntary compliance emphasis
federal funding driving programs	→	creative financing
balancing economic and environmental issues	→	integrating economic and environmental issues

federal agency leadership/oversight ➔ federal/state partnership

Acknowlegement of socio-economic considerations and differing value systems in planning and management ➔ inclusion of socio-economic considerations and differing value systems in planning and management

single media emphasis ➔ multi-media, ecosystem approach

geo-political boundaries as the basis for planning and management ➔ hydrologic boundaries as the basis for planning and management

environmental ethic ➔ sustainability ethic

These and related trends have, and will continue to have profound implications for state governments and their water resources planning and management responsibilities. State leadership will be essential in addressing the more challenging aspects of evolving federal/state relations, such as the consequences of government downsizing, fiscal constraints, changes in regulatory regimes, and other implications associated with the movement of planning and management authority from the federal level to a partnership-based approach with enhanced local government, private sector and citizen involvement.

Building the Institutional Infrastructure: The Growing Relevance of Interstate River Basin Organizations

Interstate river basin organizations, at least in a rudimentary form, have been present throughout the history of U.S. water resources planning and management. (Donahue, 1987) An initial, exclusive focus on development for navigation purposes gradually gave way to modern institutional forms with comprehensive, multi-objective planning and management responsibilities. Today, such multi-jurisdictional institutions -ranging from legislatively mandated commissions with considerable authority to informal, voluntary associations for information exchange- are permanent fixtures on the government landscape. Further, they have emerged as a leading vehicle by which state governments can assert much-needed leadership in the "new era" of water resources planning and management. A related outgrowth is a dramatic increase in the number and type of creative intergovernmental policy statements (e.g., memoranda of understanding, charters, cooperative agreements) designed to formalize and guide emerging partnerships between and among all levels of government.

Innovations in Interstate Collaboration—The Great Lakes Experience

The binational Great Lakes Basin is in the midst of a continuing, century-old "experiment" with institutional arrangements that promote intergovernmental coordination and collaboration in water resources planning and management efforts. Two federal governments, eight states, two Canadian provinces, numerous tribal authorities and countless units of local government share management responsibility for the world's largest and most intensively used fresh water resource. Given the intensity and diversity of such use, coupled with a complex governance framework and history of innovation at the intergovernmental level, the Great Lakes Basin has understandably emerged as a case study in intergovernmental relations.

Intergovernmental coordination and collaboration in the Great Lakes Basin is accomplished, in large part, by the work of four regional organizations funded or otherwise supported by the eight state governments. A brief description follows, accompanied by one or more examples of issue-specific intergovernmental initiatives where state governments have, or will have a significant role:

The Council of Great Lakes Governors, formed in 1982 as a private, non-profit entity, provides a forum for the eight Great Lakes state governors (and their provincial counterparts) to identify, discuss and formulate policy on regional economic and environmental issues of shared interest. Among many other interstate initiatives, the Council authored the **Great Lakes Charter of 1985**, a non-binding "good-faith" agreement to strengthen Basin water resource planning and management efforts, and establish a mechanism to address large scale diversion and consumptive use proposals. In the late 1980s, the governors established the **Great Lakes Protection Fund**, a $100 million state-capitalized fund to support the regional research and management priorities of the states. More recently, the Council played a prominent role in the **Great Lakes Water Quality Initiative**, a U.S. Environmental Agency/Great Lakes states effort to promote interjurisdictional consistency in water quality standards. The Great Lakes states also successfully developed, in 1996, a **Memorandum of Understanding on the Lake Michigan Diversion at Chicago**. Working together, the Great Lakes states and the U.S. Department of Justice were able to enter into a non-binding mediation process that holds promise of resolving a century-old legal dispute without United States Supreme Court intervention.

The Great Lakes Fishery Commission, a binational organization established through the U.S.-Canada Convention on Great Lakes Fisheries in 1955, is responsible for developing coordinated research programs and ensuring the sustained productivity of the fishery. It's **Strategic Great Lakes Fishery Management Plan**, with federal, state and tribal governments as signatories, has been a well-documented success story in intergovernmental cooperation.

The International Joint Commission, a U.S.-Canada organization established via international treaty in 1909, addresses water quantity and quality disputes arising along the common frontier. The **U.S.-Canada Great Lakes Water Quality Agreement**, originally signed in 1972, provides a framework for federal/state/provincial coordination in meeting water quality goals. In November 1997, the Commission proposed that a series of ten binational river basin councils be established along the entire U.S.-Canada frontier as a means to enhance intergovernmental coordination on an array of water resources planning and management initiatives.

The Great Lakes Commission is an interstate compact agency founded in 1954 via state and federal legislation. Its information sharing, policy coordination and advocacy functions are applied to an array of resource management, environmental protection, economic development and transportation issues of regional significance. The Commission brokered the **Declaration of Indiana**, a 1991 binational maritime agreement that brought diverse sectors of the maritime industry together to develop and implement a common agenda. Three years later, the **Ecosystem Charter for the Great Lakes- St. Lawrence Basin** was developed, a landmark initiative presenting an extensive series of principles and associated objectives for Basin management that has attracted endorsements from over 170 public agencies and private sector organizations. The Commission-staffed **Great Lakes Dredging Team** is providing a forum for federal/state partnership in resolving dredging and dredge disposal issues, and its **Great Lakes Panel on Aquatic Nuisance Species** is providing public agencies, utilities and commercial interests with a mechanism for basinwide prevention and control initiatives. Model state legislation and a regional action plan are presently being developed.

As noted earlier, sustainable development has emerged as a common theme and interest at all levels of government in recent years. Broadly defined, sustainable development reflects a level of resource stewardship and economic activity that provides for the needs of the current generation without compromising society's ability to meet the needs of future generations. Principles of sustainable development reflect the ecosystem-based philosophy of the Great Lakes states, and implementation demands an intergovernmental approach and public/private sector partnership. Thus, intergovernmental organizations in the Great Lakes Basin, such as those identified above, are ideally suited to move these principles from concept to application. The Great Lakes Commission, for example, sponsored an October 1997 forum at which government, industry, citizen and academic leaders discussed obstacles and opportunities in incorporating such principles into public and industrial policy. Brownfields redevelopment—the "recycling" of former industrial sites for beneficial economic use—is one practical example of sustainable development that is attracting attention throughout the Great Lakes Basin. A Council of Great Lakes Governors' initiative, with Great Lakes Commission support, is providing a forum for interstate exchange of ideas on redevelopment approaches and opportunities for joint action.

Several of the Great Lakes states share a portion of other river basins where similar intergovernmental initiatives demonstrate the emerging leadership role of state governments in addressing water resources problems and opportunities. For example, states in the Upper Mississippi River Basin are partnering with the U.S. Army Corps of Engineers on a navigation study to facilitate improvements to the Upper Mississippi and Illinois Waterways, and are also engaged in an environmental management program with the Corps and the U.S. Department of the Interior. Further, subsequent to the 1993 flood, these states partnered with states in the Lower Missouri River Basin to develop shared flood plain management principles. Initiatives similar to these are increasingly found in regions throughout the nation as states assert their leadership—both individually and collectively—on the problems and opportunities of water resources planning and management.

Closing Statement: A Blueprint for the Future

Strengthening and sustaining state leadership in the "new era" of water resources planning and management is critical to the future of the resource and those who depend upon it. Toward that end, several challenges must be embraced. First, state governments must fully recognize that stewardship responsibility rests primarily with them; the prescriptive, "top-down" approach to planning and management is indeed a thing of the past. Second, states must take full advantage of their regional, multi-jurisdictional basin organizations—both current and prospective ones. The need for their information sharing, coordination and advocacy services has never been greater, and their potential remains largely untapped. Third, partnerships at all levels—and in all forms—must be pursued in an era where resource constraints demand an unprecedented level of efficiency and effectiveness in program design and implementation. And, finally, we must continue our slow but steady progress in embracing and applying principles of sustainable development, recognizing that water resources planning and management has far-reaching environmental, economic and social implications for both the current and future generations.

References

Donahue, Michael J. 1996. "A New Era for Regional Water Resources Management: A Great Lakes Case Study." The 1996 Wayne S. Nichols Memorial Lecture, Nov. 14, 1996. The Ohio State University, Columbus, Ohio.

Donahue, Michael J. 1987. Institutional Arrangements for Great Lakes Management: Past Practices and Future Alternatives. Michigan Sea Grant College Program. Ann Arbor, MI. 394 pp.

Great Lakes Commission and Federal Reserve Bank of Chicago. 1990. The Great Lakes Economy: Looking North and South. Federal Reserve Bank of Chicago, Chicago, IL. 163 pp.

COLUMBIA AND SNAKE RIVERS

Moderator, Eric Strecker

"The Corps Perspective on Salmon Recovery Efforts"*
Greg Graham
Corps of Engineers

"The Columbia River's Hydropower, Salmon and Water Policy"
Jack Wong
Northwest Power Planning Council

"Improving Coordination from an Environmentalist Viewpoint"*
Lori Bodi
American Rivers Council

"A Tribal Perspective on Environmental Issues"*
Rob Lowthrope†
Columbia River Intertribal Fish Commission

*Text not available at time of printing.
†Attendance not assured.

The Columbia River: Hydropower, Salmon and Water Policy

Jack Wong[1]

Abstract

In the Pacific Northwest, difficult environmental, scientific and economic issues confront planning efforts for the Columbia River, a river that is largely rural, but also the economic engine for a region of 9 million people through its hydropower generation. It is a river with longstanding conflicts of purpose: an industrial river on one hand, and a river that continues to sustain fish and wildlife on the other. It is a river that died and was reborn as money, according to one writer, and also once was the world's greatest salmon river.

The tension between these two characteristics of the Columbia, and our region's efforts to restore fish and wildlife while continuing to provide economic benefits from the river and its dams, is the focus of this paper.

The Physical Environment

To understand the conflicts on the Columbia, it helps to understand its physical environment. The headwaters of the Columbia are in southeastern British Columbia. The river is 1,214 miles long. It carries ten times as much water as the Colorado River and 2.5 times as much as the Nile. It is the fourth-largest river, by volume, in North America, annually emptying some 180 million acre feet of water into the Pacific Ocean near Astoria, Oregon. The Columbia estuary, where all of this water mixes with salt water, covers about 95,000 acres – 150 square miles.

The southern rim of the basin runs through northwest Wyoming, southeast Idaho, the corner of Utah, small parts of northern Nevada and most of Oregon. The northern rim runs from Wyoming along the Idaho-Montana border, cuts east past Butte, Montana, heads north into Canada, and loops back south along the crest of Washington's Cascade Mountains, then west across southwestern Washington to the ocean. The basin covers 259,000 square miles, roughly the size of France.

The amount of annual runoff varies dramatically around the basin, from less than 12 inches of precipitation annually in the southern basin to 100 inches or more in the Cascade Mountains on the west.

---

[1] Director, Fish and Wildlife Division, Northwest Power Planning Council, 851 S.W. Sixth Avenue, Suite 1100, Portland, Oregon, 97204. The author wishes to acknowledge the substantial contribution of John M. Volkman, General Counsel of the Northwest Power Planning Council, to this paper.

The geography of the Columbia Basin includes a vast subterranean component, as well. The interaction of ground water and surface water is incompletely mapped, but critical to water management. For example, in the middle Snake River, the relationships of surface flows, aquifer discharge, flood irrigation, hydropower generation, groundwater pumping, and new irrigation efficiencies are crucial considerations in water allocations.

Cultural Communities

The Basin also is diverse culturally. The far reaches of the basin are sparsely populated, but economically diverse. Jackson Hole, Wyoming, where the billionaires are said to be squeezing out the millionaires, is in the basin. So is the virtually empty basin of the Malheur River, in the desert of southeastern Oregon and Northwestern Nevada, where gold miners once roasted salmon over greasewood fires. Fairmount Hot Springs, in a stunningly beautiful valley at the base of the Canadian Rockies, in the basin. So is the metropolitan area of Portland, Oregon, and Vancouver, Washington, with a combined population of 1.2 million.

The basin apparently has been occupied by humans for at least 12,000 years. Tribal groups on the Columbia Plateau hunted, fished, trapped and gathered to sustain themselves. They fished for salmon, steelhead, sturgeon, trout and other species at numerous places along the river and its tributaries. Following the Lewis and Clark Expedition of 1804-1806, Euro-Americans began to populate the basin as fur traders, trappers, missionaries, homesteaders, farmers, miners, ranchers, and loggers. Today, the population of the basin is about 5 million, or roughly 53 percent of the Pacific Northwest population. Much of the population is concentrated in the developed urban areas, which account for less than 4 percent of the United States portion of the basin.

Natural Heritage

When Lewis and Clark first encountered the Columbia, in October 1805, they described a river full of salmon. Fall chinook would have been returning to spawn, and the explorers remarked on the vast number of these fish and the clarity of the water. Crowds of Indians were gathering fish from the river. We estimate that prior to about 1850, when rapid development of the Pacific Northwest commenced, there were between 10 million and 16 million adult salmon and steelhead returning to the Columbia River each year.

Salmon and steelhead are born in creeks, rivers and lakes and migrate downstream to the Pacific Ocean as juveniles, grow in the ocean and then return to their birth places to spawn. Salmon of the Columbia River Basin are viewed as regional, even national treasures, and are part of the Basin's natural endowment. The Columbia Gorge National Scenic Area, the Hells Canyon National Recreation Area, and a variety of wild and scenic rivers all are within the Columbia River Basin. All are affected by water policies developed by federal stewards.

Government's Presence in the Basin

While states issue water rights and manage fish and wildlife under state law, the federal government's influence over the mainstem of the Columbia is pervasive. Various agencies of the federal government negotiate and carry out obligations under treaties with Canada related to hydropower production and fish harvest in the basin, and also treaty rights to fish reserved by Columbia River Indian tribes more than 140 years ago.

The federal government operates and markets hydropower from 14 large-scale dams on the mainstem of the river, licenses non-federal dams in the basin and operates dozens of water storage projects that irrigate more than 3 million acres. Federal agencies also manage lands that account for 55 percent of the basin. The federal government administers the Endangered Species Act in the basin, most notably on behalf of Snake River salmon, and Congress funds, or declines to fund, a number of efforts that have implications for how water is used in the basin.

Indian tribes, states, counties and cities also exercise various authorities over the river and its water, and the result is a milieu of governance that, graphically depicted, might look like a mass of cooked spaghetti. Two legal scholars, writing in the early 1980s, followed a mythical salmon from its birth in a tributary of the Clearwater River in central Idaho. To reach the ocean, the fish traveled through 17 separate legal jurisdictions.

The rapid development and industrialization of the river, the destruction of spawning and rearing habitat, the construction of the hydropower system -- it generates 9,000 megawatts of energy and $2 billion in annual income, -- and myriad other human-caused impacts led to the steady decline of salmon and steelhead. From historic annual returns of as many as 16 million fish, the runs declined to something like 3 million today -- and most of those are hatchery fish. No wonder historian Richard White described the Columbia as "The Organic Machine," and historian Donald Wooster described it as "a river that died and was reborn as money."

Fragmentation of River Uses -- Past, Present and Future

So the Columbia River's problems today result from its remarkable, but contradictory, attributes -- a once-prolific salmon river on one hand, an artery of commerce and industry on the other. It may be an oversimplification, but it is instructional to briefly examine the river's fragmented uses as context for our future water policy choices.

First, a look at the past. Put simply, the river was exploited and the salmon declined. The Columbia always has been seen as a river of commerce. Thomas Jefferson instructed Lewis and Clark that the object of their mission was to search out a water route across the continent for purposes of commerce. The lucrative fur trade between the Columbia River and China, in which profits of 3,000 percent were not uncommon, began almost immediately after the river was discovered in 1792. The basin's vast forests, mineral resources and, of course, salmon, attracted fishermen and fish-packers, loggers, miners and farmers. Cities were born, and many flourished. Thousands of people traveled the perilous Oregon Trail from Missouri, beginning in the mid-1840s. Irrigated agriculture began around 1850, there was a gold rush in Idaho in the 1860s. Steamboats plied the Columbia and Snake rivers as far inland as Lewiston, Idaho -- 450 miles from the sea. The transcontinental railroad arrived in the 1880s. And so it went.

Numerous canneries pumped out millions of cases of 48 1-pound cans of salmon, enhancing the region's stature and allure in places like New York and London. With the discovery of electricity, and later the development of hydroelectricity, the region's potential glowed, literally.

By the 1920s, the federal government was investigating hydroelectric potential in the Columbia River Basin. The first dam to span the mainstem of the Columbia was completed in 1933. It was Rock Island, built by a private power company. Construction started that same year on the first two federal dams, Grand Coulee and Bonneville. The last of the big dams in the federal power system, Lower Granite on the Snake River, was completed in 1975.

As the Pacific Northwest grew and the Columbia was exploited for its water, fish, hydropower and navigation, salmon runs steadily declined. There are a number of reasons. The impact of the hydropower dams may be the most significant, as the dams were impediments to downstream and upstream salmon migration. But overfishing also was a problem -- Columbia River salmon were fished nearly to extinction by 1900. Another key problem was degradation of spawning and rearing habitat by logging and agriculture, which promoted sedimentation, turbidity, and lethally high stream temperatures in many important salmon-spawning streams.

Part of the reason for creating the Northwest Power Planning Council in 1980 was to address these problems. The law that created the Council requires us to plan for the region's future sources of electricity while also protecting and enhancing the fish and wildlife that have been affected by the dams, which are the region's largest single power supply. Since 1981, when the Council adopted its first fish and wildlife program, we have been working to improve conditions for fish while also assuring the region of an adequate electric power supply.

Today, this remains one of our biggest dilemmas: How to enhance fish and wildlife populations while continuing to operate the dams for their multiple purposes. I should make clear that the Power Planning Council does not operate the dams, but the federal agencies that do -- Bonneville, the Corps of Engineers and the Bureau of Reclamation -- are required by law to take our fish and wildlife program into account when making river-operation decisions.

Managing the river for fish *and* power production is challenging, made all the more complex because river management also is an exercise in international relations.

The Columbia River Treaty of 1964 authorized the construction of four large storage dams in the upper Columbia River Basin. Three are in Canada, and the fourth is in Montana. These dams take advantage of the fact that most of the

Columbia's water originates in the headwaters region. In fact, 44 percent of the water in the river east of the Cascades originates in British Columbia.

These storage dams help regulate the Columbia's flows, which historically were high in the spring and low in the winter. This evening-out allows river operating agencies to maximize power generation by operating the entire hydropower system in a coordinated way.

The Columbia River Treaty addresses only flood control and power generation. Other water uses were not addressed, and there is no formal way to incorporate the needs of salmon or other environmental considerations into river management through the treaty other than through economic trades. However, the needs of endangered Snake River salmon are addressed under the Endangered Species Act by the National Marine Fisheries Service, which directs river operations on the Snake and lower Columbia rivers. Of course, these river operations generate controversy, particularly with environmental groups.

Because most salmon runs continue to decline, it seems clear river operations will have to change in the future, in concert with continued efforts improve salmon survival at other stages of the life cycle.

Here are several observations about the future:

First, the Corps of Engineers likely will decide in the year 2000 whether to breach the four lower Snake River dams or to enhance barge transportation of Snake River salmon smolts past the dams. Obviously, there are huge implications for the salmon, and for the hydropower system. Breaching, which is tantamount to dam removal, would give the salmon a free-flowing river but eliminate power generation at the four dams, which provide about 5 percent of the region's electricity.

Second, a recent scientific report calls for a re-examination of current approaches to salmon recovery. This report, entitled *Return to the River,* was written by a panel of nine independent scientists. The report portrays two competing views of the Columbia as a working river. One view is of an industrial river of hydropower generation, irrigation and flood control, and fish species that migrate in time periods that fit with harvest plans and minimize conflicts with hydropower generation. The other view is of a more complex and natural river, one that works through natural functions to support a rich diversity of species and a bountiful food chain. *Return to the River* urges that salmon recovery be premised on the restoration of a working salmon ecosystem -- a collection of healthy salmon habitats connected by healthy rivers. Whether this is possible in the industrial Columbia is a political judgment, not a scientific one.

Third, the rapid deregulation of the nation's electricity industry is forcing the Columbia River hydropower system into competition with other power suppliers -- a thing that would have been unimaginable just 10 years ago. There are implications for salmon in this competition, in that the federal hydropower system pays for most of the fish and wildlife restoration effort in the Columbia River Basin. Currently, Bonneville's budget for direct-funding of fish and wildlife projects is capped at $127 million per year. This agreement lasts only through 2001 -- coincidentally the same year that 90 percent of Bonneville's power sales contracts expire. Recently, a panel of experts in energy policy and corporate finance recommended a number of cost-cutting measures designed to lower the cost of Bonneville's power to the anticipated market price -- or lower -- by 2001. This would help Bonneville return to its historic mission of selling federal hydropower at cost, primarily to public utilities. The direct-spending portion of Bonneville's fish and wildlife budget could remain at $127 million, or it could rise -- particularly if Bonneville is required to pay for dam-breaching or enhanced barge transportation of Snake River juvenile salmon.

Policy issues and recommendations

Finally today, I'd like to discuss four water policy issues that arise from our experience with the Columbia.

First, it seems clear that water management initiatives have to fit into a much broader series of recovery measures that, collectively, add up to healthy ecosystem conditions. But how do we define those healthy conditions? How do we measure the contribution of individual initiatives, and how do we link them analytically with a large collection of other measures to determine whether they add up, or even if we have a solvable problem? The industrial river is in place; science is murky, but nonetheless persuasive. We have to make judgments about tradeoffs and risks based on a body of knowledge that is, at best, emerging.

Second, when water management is as fractured as it is in the Columbia River Basin -- federal, state, tribal, Canadian and private ownerships -- how can we achieve the degree of coherence that an ecosystem approach to water policy suggests? The Endangered Species Act has helped to pull federal agencies and programs together, but there are many unconnected pieces -- tributary watersheds, for example, where Western water rights have a firm grip, and the Canadian portion of the basin, where our federal laws have no impact.

Third, it seems clear that the balance between hydropower and salmon recovery, between the river of industry and the river of salmon, has to be re-established. Hydropower's financial contribution to salmon recovery has to be put on more stable footing. Improved accountability in ecosystem recovery programs is needed to ensure that limited funds are spent wisely and are acceptable to the public. And the financial basis for ecosystem recovery must be broadened beyond the hydropower system.

Finally, a new paradigm is needed for river governance. The management of the federal hydropower system is inextricably linked to salmon-recovery decisions, and *vice-versa*, and yet energy policy is increasingly driven by market forces and decreasingly by environmental considerations. Thus, we need better-integrated government institutions that can mandate collaboration, establish consistent government policy and manage dams and recovery efforts in ways that complement each other.

COLORADO RIVER

Moderator, Howard Holme, Esq.

"How Can We Improve Coordination of Water Resources Planning with the Endangered Species Act?"
Howard Holme
Fairfield and Woods

"Improved Coordination of the ESA Activities: How We Can Help Each other?"
Ralph O. Morgenweck
U.S. Fish and Wildlife Service

"Why the Colorado River Endangered Fish Are Not Recovered, and What Needs to be Done"*
Tom Pitts
Hall Pitts and Associates

"Can a Conservation Plan Save 102 Species at Reasonable Cost?"
Gerald Zimmerman
Colorado River Board of California

*Text not available at time of printing.

How can we 'improve coordination' of water resource planning with the Endangered Species Act?

Howard Holme[+]

Abstract

After a brief introduction to the Colorado River, this paper advises people how to coordinate water resources planning with the Endangered Species Act. Use state or private programs to avoid listing of species. Use scientists to resist bad science and alarmism. Recognize the power of the Act to protect Threatened or Endangered (T&E) species. Try to cooperate with the federal agencies. Use sections 9 and 10 of the ESA, emphasizing Habitat Conservation Plans and incidental take permits. Avoid section 7(a)(2) of the ESA. It requires federal agencies to insure actions almost exclusively for the benefit of T&E species. Use section 7(a)(1) to require live capture, transplantation, stocking of endangered species, and removal of nonnatives, predators, and competitors to T&E species. Use market mechanisms to recover species. Amend the Act. Provide a federal program to buy habitat and provide money for needed condemnations and inverse condemnations. Amend the Act to emphasize cost-effective preservation of genes, and to a lesser degree, species.

Introduction to the Colorado River

The fundamental reason environmental laws need to be coordinated with, and sometimes subordinated to, water resources planning in the Colorado River Basin is that the basin has a substantial population, but little water. Water resource planning is extremely important in the Colorado River Basin. This basin includes

---

[+]Howard Holme, Stanford University A.B., 1967 with distinction, and Honors; Yale Law School, J.D. 1972, is an environmental and water lawyer, and President of Fairfield and Woods, P.C., Suite 2400, 1700 Lincoln St., Denver, CO 80203-4524. (303) 830-2400, fax (303) 830-1033. hholme@fwlaw.com.

about 8% of the land area of the United States, but receives less precipitation than the Delaware River that drains less than a tenth as much land.[1] The Colorado River and its tributaries serve about one-fifth of the national population[2] with so little precipitation that most of the land is a desert.

For the last 100 years, the federal, state and local governments have worked hard to make the desert bloom and grow billions of dollars of the nation's food and fiber, using about three quarters of the water that is consumptively used.[3] The US Bureau of Reclamation (U.S.B.R.) dams, aqueducts and projects are extremely important. The US built Hoover Dam to store 27.4 million acre-feet (MAF) in Lake Mead, Glen Canyon Dam to store 25 MAF in Lake Powell, and other dams to store a total of 62 MAF.[4] The Central Valley Project, the Westlands Project, the Imperial Irrigation Project, the Central Arizona Project, and the Central Utah Project are some of the better known projects.

Water is apportioned to states under Compacts and the "Law of the River." Colorado contributes 71% of the water, and is apportioned 24% of the water. Arizona contributes 1% and is apportioned 20%. California contributes almost nothing, and is apportioned 30%.[5] California uses more than its 4.4 Million Acre-Feet of water per Year (MAFY), Nevada wants more than its 0.3 MAFY allocation, Arizona is now drawing its allocation of 2.8 MAFY, and Utah, Colorado and Wyoming want to be able to develop their Colorado River Compact and Upper Colorado Compact shares.[6] Average total river flows are about 14 MAFY.[7]

A large population, of environmentally sensitive people, in a very arid area, inevitably feel conflicts in their values. Environmentalists argue as follows. Strengthen environmental laws. Social, governmental, and environmental stakeholders should partner in watershed governance. Reduce human consumption to maintain sustainable uses of water for Native Americans and for rare species. Avoid diversions. Conserve water unpolluted for instream flows, recreation, and restoration of wetlands and aquatic ecosystems. Remove dams. Recreate floods. Prevent pollution. The public trust and public interest outweigh outmoded notions of private property.[8]

Water agencies that provide water to the growing population seek to balance environmental values against practicalities of costs and benefits, demands to maintain water as the second cheapest commodity next to air, maintenance of existing water diversions and historic taxpayer investments. Water providers see growing populations of humans who do not want to be thirsty, hungry, or to have expensive cotton or brown lawns. Providers see shrinking federal government subsidies and growing federal regulations and inhibitions, from the Bureau of Reclamation to the Environmental Protection Agency.

Most Colorado River Basin water specialists are spending an enormous amount of their time and effort dealing with one federal obstacle to water operations, the Endangered Species Act. The Upper Basin has the Recovery Implementation Plan Recovery Action Plan (RIPRAP) for four Endangered Colorado River fish, the Colorado Squawfish, the Humpback Chub, the Razorback Sucker, and the Bonytail Chub. The Lower Basin has these four fish, and is planning its Lower Colorado River Multi-Species Conservation Program (LCR MSCP) for another 98 species. More than twenty native western fishes have become extinct in the past century, and one hundred more are considered threatened, endangered, or of special concern.[9]

On one side, environmentalists say the Act has not prevented use of historic facilities, that unavoidable jeopardy has been avoided, and that Habitat Conservation Plans, Safe Harbors, No Surprise Agreements, and Reasonable and Prudent Alternatives are sufficient.[10] On the other, water providers are extremely frustrated at endless meetings for slow moving bureaucratic efforts to recover species, at demands for "all remaining water in the streams" except for minor "carveouts," and by feeling existing facilities and future growth are continuously in jeopardy.[11]

We turn now to hints on how water providers might cope with the ESA, recognizing that none of them may suffice.

1. <u>Use state and private programs to avoid listing of species.</u>

An ounce of prevention is worth a pound of cure. Usually there is a warning before a species is federally listed. The species is mentioned in the newspaper, in environmental newsletters, in suits demanding listing, or in Federal Register Notices. Private parties or programs and state wildlife agencies sometimes recover species before they are listed, at a minor fraction of the expense and trouble if the species is federally listed. In Colorado, the Greenback Cutthroat Trout has been sensibly handled, the boreal toad has avoided federal listing, and many other species may be recovered by state and private programs. Once a species is federally listed, it is generally illegal for private parties to collect endangered animals, grow them, or recover them.

2. <u>Use scientists to resist bad science and alarmism.</u>

Scientists are useful in some, but not all, ESA issues. Can they help say a proposed "species" is not a species? Usually not. The preservation of species is a less important motivation to many ESA supporters than land or "habitat" preservation and prevention of a particular economic development. Further, "species" was defined by politicians and not scientists. The dictionary defines "Species," as "a group of intimately related and physically similar organisms that actually or potentially interbreed and are less commonly capable of fertile interbreeding with members of other groups . . . "[12] Many or most federally listed species would not satisfy this definition. Scientists are often *not* useful in resisting listing under the ESA definition including "any **subspecies** of fish or wildlife or plants, and **any distinct population segment** of any species of vertebrate fish or wildlife which interbreeds when mature."[13] Can an "endangered species" be as small as a monogamous pair of birds, because each is a "distinct population segment . . . which interbreeds when mature?"

Scientists, engaged early, should be useful in controlling the amount of habitat that is listed as critical habitat. Water agencies did not use scientists to resist the critical habitat listing, but 16 U. S. C. § 1532 (5) defines "critical habitat" as "the specific areas within the geographical area occupied by the species, **at the time it is listed**." The four Colorado River Endangered Fish have critical habit listings which extend 1,000 miles from California throughout the Colorado River Basin high into Colorado and Utah. However, at the time the fish were listed, they existed only in a few isolated locations along the river and its tributaries.[14] As far as we could tell, the best excuse for listing the bonytail chub habitat in Colorado was one, single, potential sighting.

Scientists could also prevent the mass stocking of nonnative predators of the endangered species in the critical habitat of the endangered species. They should be helpful in the encouragement of transplantation of endangered species rather than the federally encouraged euthanasia of endangered species. They might be useful in discouraging the mass listing of species. They might have some effect in resisting, for example, the efforts of Jasper Carlton, who is reported to have personally caused the listing of 340 of the 944 species covered by the Endangered Species Act in 1995.[15] Scientists might be helpful in arguing that we should amend the Act to avoid some its bad results.

3. <u>Recognize the power of the Act to protect Threatened or Endangered (T&E) species. Try to cooperate with the federal agencies.</u>

Because of the stringency of the ESA and its primacy over agency primary responsibilities, we can accomplish only a limited amount of "coordination" with the ESA. The first Supreme Court interpretation of the Endangered Species Act (ESA) gave a literal reading to the extremely stringent words in the Act.[16] The Court upheld an injunction to prevent imminent completion of the Tellico Dam that would probably jeopardize the 15,000 snail darters. The Court said the following. "Congress has spoken in the plainest of words, making it abundantly clear that the balance has been struck in favor of affording endangered species the highest of priorities."[17] Endangered species take a higher priority than agencies' primary objectives or ongoing projects.[18] "Various authorities calculate . . . some 200,000 [species]--may need to be listed as Endangered or Threatened."[19] The plain intent of Congress in enacting [the Act] was to halt and reverse the trend toward species extinction, whatever the cost. . . ."

Agencies comply with the ESA first, and only secondarily consider the Clean Water Act, the National Environmental Policy Act, or the Reclamation laws. Compliance with section 7 of the ESA is so difficult that it frequently "trumps" the other acts, and frequently postpones or prevents actions that other acts would allow. Coordination of water resources planning and development with the ESA often turns into a struggle to comply with the ESA.

One course of action is to wait for the application of ESA section 7 (a) (2).[20] Section 7 requires the Secretary to "**insure** that **any** action authorized, funded, or carried out by such agency . . . is **not likely to jeopardize** the continued existence of any endangered species or threatened species or result in the destruction or adverse modification of habitat of such species which is determined . . . to be critical." The probable result of allowing full section 7 review and action is delay, bureaucratic control, and huge expense for water agencies.

A better course of action is to cooperate with the federal agencies when possible. The U.S.F.W.S. has enormous power and discretion. The Upper Colorado River Basin used Section 7 procedures and will spend $100 million in 20 years to try to recover four fish, facing considerable resistance from the government. In contrast, the Lower Colorado River Basin worked out an agreement to spend $4.5 million by December 2000 to develop the fifty-year Multi-Species Plan, save more than 100 species, and is being praised by Secretary Bruce Babbitt.

Courts give great deference to U.S.F.W.S. findings, even when the agency reverses its course to favor people over fish.[21] One suit sought an injunction to require the release of about four million acre-feet of water from Lake Mead, an amount nearly equal to California's total annual Colorado River share, in order to preserve some newly grown willows in which a few pairs of birds nested. The opinion, dismissing the injunction to lower Lake Mead, relied heavily on DOI expertise, allowed a "tiered approach to reach a no jeopardy determination" even in light of limited scientific information, allowed the Service not to protect the second largest willow grove, and allowed potential significant, if not complete mortality to the Lake Mead habitat and Lake Mead Flycatcher population.[22] The Court also relied heavily on the lack of discretion of the DOI to deliver Colorado River water in accord with the Law of the River.[23] The case is being appealed.

2. Use sections 9 and 10 of the ESA, emphasizing Habitat Conservation Plans and incidental take permits. Avoid section 7(a)(2) of the ESA. It requires federal agencies to insure actions almost exclusively for the benefit of T&E species.

Section 9(a)(1) says that except with a permit it is unlawful for any person subject to the jurisdiction of the United States to . . . take any endangered species of fish or wildlife.[24] Individual animals are protected from direct or indirect effects which might harm, harass or kill them.[25] A citizen's suit may be brought to enjoin road construction and timbering within one-half mile of the previous year's nest of a single pair of Northern Spotted Owls.[26] Almost no water resource activity can insure it will not cause a taking. Water diversion, or water storage, or water treatment, or water discharge will change the habitat, and can be argued to cause a taking. When endangered animals are protected, one by one, taking is practically inevitable, and taking is unlawful except with a permit. Thus, a takings permit is practically required.

Section 10 of the ESA, 16 U.S.C. § 1539 allows the US Fish and Wildlife Service (U.S.F.W.S.) to permit, under certain conditions, any taking otherwise prohibited by Section 9 if such taking is incidental to carrying out an otherwise lawful activity. The applicant must propose a conservation plan convince agency that proposed action will minimize and mitigate the impacts of the taking and will not appreciably reduce the likelihood of the survival and recovery of the species in the wild. The Draft Habitat Conservation Planning Handbook issued by the U.S.F.W.S. provides detailed steps for establishment of Habitat Conservation Plans (HCPs).

6. Use section 7(a)(1) to require live capture, transplantation, stocking of endangered species, and removal of nonnatives, predators, and competitors to T&E species.

Water agencies and providers should demand that the U.S.F.W.S. and other federal agencies take conservation actions. Under section 7(a)(1), federal agencies "**shall . . . use . . . all methods** and procedures which are necessary to [recover and delist] any endangered species or threatened species. . . .[including] **habitat acquisition and maintenance, propagation, live trapping, and transplantation** . . . "

The only light at the end of the ESA tunnel is recovery of the species. Yet federal agencies may be very reluctant to stock endangered fish.[27] For example, almost no bonytail chub have been found in the Upper Colorado River except very near Lake Powell in many years.[28] Reports say the fish is essentially extirpated in the Upper basin and the only means of recovering bonytail chub is through hatchery growth and reintroduction. [29] The 1987 RIPRAP properly planned that "The bonytail will be reintroduced immediately into the upper basin to improve the status of the species and to provide adequate numbers to study habitat needs."[30] We think only very small numbers of fish have been stocked. The excuse for not stocking was potential inbreeding. Inbreeding is a poor excuse if only a few dozen fish are known to exist.[31] The failure to do augmentation and restoration stocking of bonytail appears to violate the Endangered Species Act.[32]

The 1987 RIPRAP recognized that stocking should be done.[33] It proposed to raise 1,000,000 squawfish for the first five years of the program. It proposed 500,000 razorback, 200,000 bonytail, and 150,000 humpback chub be hatched per year.[34] Only tiny fractions of these numbers of fish have been stocked. Instead, the agency planned for research rather than stocking, planning to "thin" and sometimes to "**euthanize**" native fish.[35]

6. Use "experimental populations" and market mechanisms to recover species.

Endangered animals in official or nonexperimental populations of endangered species cannot be "taken." Too often this means that heroic efforts need to be undertaken to improve habitat, when the real problem is nonnative predators or competitors. The endangered species may be a poor competitor, and nonnatives may be difficult to remove without taking endangered animals. Under section 9(a) and 10(j) of the Act, populations can be designated "experimental," released "wholly separate geographically from nonexperimental populations," exempted from takings prohibitions, and allowed to grow much more rapidly than official populations typically grow. New geographic areas for the endangered species can be cleared of predators and competitors before the translocation, endangered squawfish can be grown to a size where they can eat the red shiners and not vice versa, chub can become the predators and not the prey, etc.

People could recover reproductively prolific species quickly with the right incentives. Single adult fish can produce millions of eggs. Provide fish eggs and pay $20 per adult fish. Encourage private or state hatcheries instead of preventing them. Stop stocking predators; the U.S.F.W.S. and Colorado Department of Wildlife have stocked predatory trout in the critical habitat of endangered fish. Remove creel limits on predator fish. Pay $10 per pound for dead predator fish. Pay people to grow endangered fish, rather than trout in ponds. Plant fish above dams and diversion structures before, or instead of, building million dollar fish ladders used by one--count it, one--endangered fish the first year.

7. Amend the Act. Provide a federal program to buy habitat and provide money for needed condemnations and inverse condemnations. Amend the

Act to emphasize cost-effective preservation of genes, and to a lesser degree, species.

Congress did not pass the ESA as a land use law to prevent unwanted economic and land development. However, largely, that is its present purpose, and many people support that purpose. The Congress should pass a law for that purpose, and admit the United States should pay landowners for the uses landowners lose while the habitat is being conserved. Such a law 1) would be more honest, 2) would be more versatile in allowing condemnation for uses other than species, 3) would put the expenses within the budget where they belong, and 4) would pay landowners the losses they currently suffer without compensation.

The ESA is often justified based on saving species that may have unique or rare genes which might cure diseases or provide major benefits. We could serve that very important purpose for a fraction of the cost. Seed banks, species collections from rainforests and foreign countries, habitat purchase and conservation easements in foreign countries with species rich ecosystems, soil collections, and specific DNA and gene collections would all preserve vital genes at a minuscule fraction of the cost of preserving genes by the current ESA.

Recent research shows that the 1.3 million described species (over half of which are insects) are "just a smidgen of life's genetic diversity," found in "untold millions of individual species . . . still . . . hidden in the microbial world. Genetic analysis detects "dozens of groups of bacteria, archaea, and single-celled eukarya that are at least as genetically distinct from each other as animals are from plants, enough to qualify them as separate kingdoms of life."[36] Millions of microbial species harbor more exotic, medically important, and industrially important genes than the genes in most of the species listed under the Act. The thermophilic microbes from geysers and ocean vents,[37] sulfur reacting species in the deep earth, and microbes living in methane environments,[38] are a few examples of microbes with genes far different from the minor variations in relatively common genes between bonytail and humpback chub. These chubs hybridize, and probably few, if any of their genes are very different form other chubs.

Yet the microbial species are not, and cannot practicably be, preserved by the current Act. "A pinch of soil can contain 1 billion microbes or more . . . 'Two microbes could be as different from each other as a grizzly bear from an oak tree, and you'd never know it.' . . . [O]nly about 1% of bacteria can be grown in culture . . . [W]hile one side of a pebble may be exposed to oxygen, the other may not, thereby fostering the grown of aerobes on one side and anerobes on the other." [39] To deal with these species, and to make the ESA more practical and cost-efficient, Congress should revisit and substantially rewrite the Endangered Species Act.

1. D. Getches, "Competing Demands for the Colorado River," 56 *U. Colo. L. Rev.* 413 (1985). *Arizona v. California*, 373 U.S. 546, 552 (1963).

2. Public Review Draft, Report of the Western Water Policy Review Advisory Commission, *Water in the West: The Challenge for the Next Century*, October 1997, ("Western Water") P. 2-35.

3. *Id.* at P. 2-45.

4. Upper Colorado River Commission, *Thirty-Sixth Annual Report* 21-27, Salt Lake City, Utah, 1984. Walter Langbein said in 1959, "Any increase in capacity [of reservoirs] will not increase the supply [of usable water in the Colorado River system]. . .There is no significant gain in net regulation between 29 and 78 million acre-feet of capacity. . . .The gain in regulation to be achieved by increasing the . . . capacity appears to be largely offset by a corresponding increase in evaporation." He said 5.5 MAF of net annual regulation was achieved from 29 MAF in capacity, and 5.8 MAF from 65 MAF in capacity. Langbein, Water Yield and Storage in the United States, *U.S. Geological Survey Circular 409*, particularly p. 4. See similarly, Harrison, Potential United States Water-Supply Developments, *J. Irr. & Drainage Div., Proc. ASCE* at 479, Sept. 1972.

5. Getches, supra, P. 418.

6. Getches, supra, P. 418.

7. Getches, supra, P. 419.

8. Western Water, *supra.*

9. W.L. Minkley, "Sustainability of Western Native Fish Resources," 63, 71 in *Aquatic Ecosystem Symposium* (Feb. 17, 1997).

10. Western Water, *supra,* P. 5-80.

11. One assessment said "instream flows in the Rio Grande, Upper Colorado, and Lower Colorado water resource regions are insufficient to meet current needs for wildlife and fish habitat, much less allow any additional offstream use." Richard W. Guldin, "An Analysis of the Water Situation in the United States 1989-2040," *USDA Forest Service Gen. Tech. Rep. RM-177*, Sep. 1989) at 16-17.

12. Webster's Third New International Dictionary of the English Language Unabridged, 1967.

13. 16 U.S.C. § 1532(16).

14. U.S.F.W.S., "Endangered and Threatened Wildlife and Plants; Determination of Critical Habitat for the Colorado River Endangered Fishes: Razorback Sucker, Colorado Squawfish, Humpback Chub, and Bonytail Chub," 59 F.R. 13376, 13378 (March 21, 1994). The U.S.F.W.S. asserted "All areas designated have **recently** documented occurrences of these fish and/**or** are **treated as occupied** habitat in section 7 consultations."

15. *High Country News* of May 15, 1995. "Carlton's application of ecosystem law reached a zenith in 1992, when **he and the Fund for Animals filed a lawsuit on behalf of 443 species** that the Fish and Wildlife Service had failed to list."

16. *Tennessee Valley Auth. v. Hill*, 437 U.S. 153 (1978).

17. 437 U.S. at 194.

18. *Id.* P. 181-185.

19. *Id.*, P. 160, Footnote 8.

20. 16 U.S.C. § 1536 (a) (2).

21. See *Southwest Center for Biological Diversity v. United States Bureau of Reclamation and Bruce Babbitt*, Order filed Aug. 25, 1997, CIV 97-0786 (U.S.D.C. Ariz.).

22. *Id.* P. 26.

23. *Id.* P. 39.

24. 16 U.S.C. § 1538.

25. 50 C.F.R. §17.3. *Babbitt v. Sweet Home Chapter of Communities for a Greater Or.*, 515 U.S. 687 (1995).

26. *Forest Conservation Council v. Rosboro Lumber Co.*, 50 F.3d 781 (9th Cir. 1995). *Pacific Lumber Co. v. Marbled Murrelet*, 117 S. Ct. 942 (1997) (habitat modification which significantly impairs breeding and sheltering of a protected species amounts to harm).

27. T. Pitts and M. J. Cook, "Draft Report: Propagation and Stocking Activities of the Recovery Implementation Program for Endangered Fish Species in the Upper Colorado River Basin: Summary, Status and Assessment," January 17, 1997, is a devastating critique of the U.S.F.W.S. stocking activities. Secretarial discretion was upheld to prevent commercial turtle breeding, despite the potential in that and other cases for commercial breeding to make the species numerous and not endangered. *Cayman Turtle Farm, Ltd. v. Andrus*, 478 F. Supp. 125 (D.D.C. 1979).

28. U.S.F.W.S., "Bonytail Chub, Revised Recovery Plan" approved by Galen L. Buterbaugh 9/4/90, and current in May 1996, P. 5-6.

29. Bonytail Chub Recovery Plan, at p. 16.

30. U.S.F.W.S., "Final Recovery Implementation Program for Endangered Fish Species in the Upper Colorado River Basin," September 29, 1987, at P. 4-12. Concerning grow out ponds see P. 4-17.

31. Despite the few fish, seven years later the 1994 Hatchery report says "the heterozygosity was comparable to mean values reported for other western North American cyprinids. Richard S. Wydoski, U.S. Fish and Wildlife Service, "Coordinated Hatchery Facility Plan: Need for Captive-Reared Endangered Fishes and Propagation Facilities," May 25, 1994, page 13. Minkley, et al. (1989) concluded that the genetic variability of the Lake Mohave broodstock as suitable for restoration stocking into appropriate natural environments."

32. See *Colorado River Water Conservation Dist. v. Andrus*, 476 F. Supp. 966 (D. Colo. 1979) which was dismissed before final determination.

33. U.S.F.W.S., "Final Recovery Implementation Program for Endangered Fish Species in the Upper Colorado River Basin, " September 29, 1987, at P. 4-12. Concerning grow out ponds see P. 4-14 to 4-15.

34. *Id.*, P. 6-33.

35. The Genetics Management Guidelines (pp.31-32) said, "Fish not identified in Items 1 through 6 will be **euthanized** following the accepted protocol of the fisheries profession . . . *Euthanasia Protocol.* Euthanasia of excess fish will follow recommendations in the "Guidelines for Use of Fishes in Field Research" that were developed jointly by the American Fisheries Society, American Society of Ichthyologists and Herpetologists, and the American Institute of Fisheries Research Biologists. J. Holt Williamson and Richard S. Wydoski, U.S. Department of the Interior Fish and Wildlife Service, "Genetics Management Guidelines, Recovery Implementation Program for Endangered Fishes in the Upper

Colorado River Basin," May 25, 1994, page 31-32. In the 1994 "Coordinated Hatchery Facility Plan" the U.S.F.W.S. plans to start with 12,500 each of razorback suckers, bonytail, humpback chub and Colorado squawfish and grow them for several years; "the fish will mature and can be 'thinned' to 60 fish per family lot (1,500 adult fish)" so the fish will take up less space.

36. R. F. Service, "Microbiologists Explore Life's Rich, Hidden Kingdoms," *Science* 1997 March 21; 275 (5307):1740.

37. "Money for Extremophiles," *Science* 1997 January 31; 275 (5300):623.

38. J. G. Ferry, "Biochemistry:Methane: Small Molecule, Big Impact," *Science* 1997 November 21; 278 (5342):1413

39. R. F. Service, "Microbiologists Explore Life's Rich, Hidden Kingdoms," *Science*1997 March 21; 275 (5307):1740-3.

Improved Coordination and Cooperation of ESA Activities: How Can We Help Each Other?

Dr. Ralph O. Morgenweck[1]

Abstract

In the late 1980's and early 1990s, implementation of the Endangered Species Act was often fraught with contention and difficulty. The ESA was often portrayed, incorrectly, as very inflexible, rigid, and diametrically opposed to legitimate resource use and development. Over the last four to five years, the Fish and Wildlife Service and Department of the Interior have instituted a number of policy and procedure provisions designed to enhance use of the real flexibility and practical interpretation inherent in the Act in order to effectively conserve endangered and threatened species in balance with the other social and economic considerations important to the American public. This presentation will illustrate several examples of how the Service and water uses have, through development of mutual trust, negotiated effective ESA recovery implementation activities that increase legal certainty for water users, including realistic cost-share arrangements, focus on fulfilling ESA requirements as opposed to avoiding them politically, incorporate provisions of State law to enhance conservation of species, enhance proactive conservation of candidate species to preclude the need to list, and use the ESA as a facilitating mechanism leading to the resolution of resource issues that are not just ESA specific. These examples are based on programs and activities already in place or being implemented at this time.

In the 1800s "Go West, Young Man" was heard over and over. And they did – and the trend continues. In all of history, no desert or semi-desert landscape has been more ambitiously altered

---

[1]Regional Director, U.S. Fish and Wildlife Service, P.O. Box 25486, Denver, Colorado 80225.

than the American West. Cities with millions of people – Los Angeles, Phoenix, even San Francisco – have grown up where the local water supply *might* have sustained tens of thousands. Immense tracts of desert have been irrigated and turned into productive farmland. It was all possible through the manipulation of water: through the erection of dams as high as skyscrapers and aqueducts that reroute rivers hundreds of miles away. –This citation from the book, *Cadillac Desert* , illustrates the history of the river I am going to talk about today, the Colorado River – a river that has been the lifeblood of the Southwest – critical to both wildlife and humans, today and tomorrow.

Until recently, and by that I mean probably within the last 15 to 20 years, everyone took from the river what they needed. I'm sure it crossed the minds of some – how long can we do this? But for many, the river was there–always had been–and no doubt would continue to be. Surely there's more water where this came from – I'm not taking that much so what's the problem. After all, any water that reached the ocean was wasted water, or so many believed. In some places, it was no longer even thought of as a river. It was water and water meant wealth to an arid land. But mankind's heroic efforts to transform nature have not been without a price. The river has changed drastically. As populations increased, so did our unquenchable thirst for water. Water for cities, for irrigation, for industry. We aren't all asking the same thing from this river, but we are all asking. And that's why it's now so important for all of us who use the river and its resources to work together in coordinating our demands for water so that this river lives on.

Problems in the Colorado River and wildlife seem to have focused around endangered fish. Not to say that everything else is running smooth as silk–but fish are where we entered the picture. The decline of fish in the arid American West is of grave concern to many. According to the book, Battle Against Extinction on Native Fish Management in the American West, the state of Arizona recognizes 27 fishes needing special protection, California 14, Colorado 15, Nevada 29, New Mexico 24, Oregon 4, Texas 30, and Utah 18. More than 88 fishes are recognized by the Federal Government as threatened or endangered in the United States alone. The federal list could even be looked upon as inadequate as the American Fisheries Society recently reviewed the status of fishes in North America and listed 254 freshwater species as endangered, threatened, or of special concern, and 214 Pacific Coast anadromous salmonid populations at risk.

How do we reconcile our goals for recovery of these fish with the often conflicting goals of creating new jobs, maintaining or improving the local economies, and providing for recreational opportunities for more and more people?

We tried to look at all the angles of the Endangered

Species Act when we first started working on the Upper Colorado Fish Recovery Program. From its inception in 1988 to present, it has a proven track record of how groups with different interests can work together to achieve common goals. Now I'm not saying this has been a thing of beauty, because sometimes it has been pretty ugly – but as they say in the art world – it's all in the eyes of the beholder. And in this case, I think all of us see progress, admittedly some less than they want – but still progress in all goals of each of the participants. This Program is a cooperative effort between Federal and State agencies, power interests, water users, and environmental groups. This cooperative effort has the goal of recovering four endangered fish species in the Upper Colorado River Basin while allowing water development to proceed in compliance with the Endangered Species Act, interstate compacts, and state water law. This Program was established because of conflicts in the Section 7 consultation arena. How could we authorize additional water depletions in biological opinions, when depletions were a major contributing factor to the fish being endangered in the first place? The purpose of Section 7 consultations was to ensure that a project's impacts didn't jeopardize the continued existence of the fish. The fish populations were depressed to levels that additional depletions could endanger the fish further. The states, water users, and Fish and Wildlife Service determined that there had to be a better way. The solution was to try and get ahead of the issue. The way to ensure future water development wasn't in conflict with the endangered fish was to make progress towards recovery with the eventual recovery of the fish as a goal. Thus the Program arose from the ashes of conflict, and became a model for addressing ESA and resource development issues.

The Program is not a "water only" program, but looks at everything that might be needed for recovery. The major elements of the Program are: 1) habitat management through protection of instream flows; 2) habitat development and maintenance, primarily restoration of floodplain habitat and levee removal; 3) managing and controlling impacts from nonnative fish species; 4) augmentation of depleted endangered fish populations through stocking; and 5) research, monitoring, and data management.

This goal of fish recovery and continued water development gave us an umbrella for working out our differences, but its success has to be measured by actual accomplishments. First, what have the water development community gotten from the Program. Since 1988, over 400 projects totaling 225,000 acre feet of water have received favorable biological opinions. These depletions represent projects in all three states covered by the Program: Colorado, Utah, and Wyoming. In 1993, I endorsed a

Section 7 agreement with the other Program participants that details how Section 7 will be administered and how implementation of recovery actions allow the Program to serve as the reasonable and prudent alternative for water depletions. The fact that we have been able to reach successful closure on so many Section 7 consultations and have an agreement in place on conducting Section 7 provides a reasonable level of certainty to the water community regarding future development.

What are the fish getting out of this cooperative effort? As part of the 1993 Section 7 agreement, a Recovery Action Plan was developed. This Plan details the actions that are needed for the recovery of the endangered fish in the Upper Colorado River Basin. Along with each recovery action is a timetable for completing the action. Based on our current scientific understanding of the fish, everything believed needed to achieve recovery is contained in the Plan. Items in the Plan are grouped into the five elements of the Program that I mentioned earlier. The plan is periodically reviewed and changes made as we obtain new information about the fish and their response to our recovery efforts.

As far as major accomplishments and hurdles, we could start with instream flow protection. It is *very* difficult to make water; therefore we had to look at how we could manage what was left in the system to meet fish and human needs. Although the fish preceded European settlement of the west, the need to provide and protect instream flows has followed much of the water development.

In 1991, the Service issued a biological opinion that detailed how water from Flaming Gorge Dam could be released to benefit the endangered fishes. The opinion called for releases to more closely mimic the natural hydrograph that existed prior to the dam. The reoperation of Flaming Gorge has resulted in higher spring releases and lower, more stable summer/fall releases. The Colorado squawfish has responded to these flow conditions. The numbers of young and adults appears to be increasing.

These water releases for the endangered fish are not made without consideration of other interests. At least twice a year, the Bureau of Reclamation and the Service hold public meetings near Flaming Gorge Dam. The purpose of these meetings is to let the public know what flows are being requested for the endangered fish and to see what flexibility there is to benefit other interests. Similarly, the Bureau of Reclamation is reoperating the Aspinal Unit on the Gunnison River in Colorado for the benefit of endangered fish.

Because of the numerous water storage projects in the Upper Colorado River Basin, spring flows have been particularly impacted. On the Colorado River, the operation of Federal and private water storage facilities is being coordinated to increase the peak and magnitude of the spring peak. The peak in 1997 was increased approximately 1,500-1,800 cfs above what would have occurred under normal operations. The Program is now looking at the feasibility of coordinated reservoir operations on the Duchesne River in Utah. Both efforts are voluntary and neither reduces the yield of respective projects. These peak flows serve to clean spawning gravels and remove sediment from the entrances of nursery areas.

Additional discretionary water is being made available from other water projects in the basin. Most of this water is being used during late summer and fall when agriculture diversions can deplete the river. Up to 10,000 acre feet of water is stored in Ruedi Reservoir for fish use; the Program is looking at whether another 21,650 cf could be provided, at least over the next 15 years. The Program and Colorado River Water Conservation District have recently completed an MOU to provide up to 6,000 af of storage water from Wolford Reservoir. All of these water projects are located on the west slope of Colorado. Many were built to provide compensatory storage for west slope development in deals worked out to allow interbasin transfer of the water for use along the east slope of Colorado, such as the Colorado Big Thompson Project and Fry-Ark Project. As we continue to negotiate solutions to meet the fishes instream flow needs, the issues that surround East versus West slope development have awakened. What started as finding solutions to fish water needs, has grown to the ESA being a vehicle to address intra-state water battles.

One of the most important actions that is being developed by the Program is called Grand Valley Water Management. Through this project the Program will fund construction of a series of checks in an existing canal system. This will result in less water being diverted to meet existing agriculture demands. The "saved water" from this project will then be available for release to an important reach of river where flows often fall below instream flow targets.

Although instream flow issues gain the most attention within the Program, we are busy implementing other "non-flow" recovery actions. Last year, construction on a fish ladder on the Gunnison River in Colorado was completed – reopening 50 miles of endangered fish habitat. During its first year,1997, 18 adult Colorado squawfish and over 15,000 native fish used the ladder. Before this ladder, fish movement had been blocked for over 75 years. Construction is underway on a fish ladder on the Colorado River near Palisade, Colorado. At least three additional ladders will be constructed. The net result will be a 60% increase of adult fish habitat in the upper Colorado River system.

Nonnative fish control efforts have been initiated throughout the Upper Colorado River Basin. An agreement was signed between the Service and the states of Colorado, Utah, and Wyoming that restricts the stocking of nonnative fish in the basin. Colorado has removed bag limits on sportfish in river reaches occupied by the endangered fish. Nonnative fish that live in streamside ponds are being eradicated and the ponds screened to prevent escape in the future. Catfish and bass are being mechanically removed from important Colorado squawfish nursery areas. Beginning next year, small minnows that are competitors and predators will be removed from nursery areas immediately prior to endangered fish spawning. Beginning in 1988, nonnative control efforts will be affecting hundreds of miles of river.

We have constructed hatcheries that are used to maintain genetic stocks of razorback sucker and bonytail. These hatcheries are now into a production mode to meet fish stocking needs, especially as we provide fish passage to historically occupied areas. Thousands of razorbacks and bonytails have been stocked in the Colorado, Green, and Gunnison Rivers.

We have a joint State/Federal monitoring program that tells us the current status of the fish, how the fish are responding to our recovery actions, and will eventually tell us when we have achieved recovery.

The Recovery Program is unique from others in that all participants must reach consensus before anything can be done. In essence, each member has veto power. This can at times result in prolonged debate and heated argument. However, each participant must realize that they are brought together by the common goal of recovering the fish.

Each of these partnerships I have described and many more across the nation are characterized by genuine trust, cooperation, mutual respect, and a desire for economic and environmental security. I believe the future of fish and wildlife conservation depends on collaborative partnerships such as these.

A quote from the author, Holmes Rolston III, puts it in perspective. If these fishes become extinct, that event alone will not stop the story underwater in the desert. Life is a many-splendored thing; fishes sparkle in desert waters. Extinction dims that luster.

SACRAMENTO, SAN JOAQUIN, AND DELTA

Moderator, Donald Du Bois

"Coordination of the CALFED Bay-Delta Program"
Steve Yaeger
CALFED Bay-Delta Program

"Optimization Modeling for River Restoration"
C. L. Lowney, Jay R. Lund, and M. L. Deas
University of California

Can a Multi-Species Conservation Plan Conserve 102 Species Including the Four Big River Fish at a Reasonable Cost

by
Gerald R. Zimmerman [1]

Abstract

The Lower Colorado River Multi-Species Conservation Program (LCR MSCP) was established in 1994 to work toward conservation of endangered and sensitive species while accommodating current and future uses of the water and power resources of the Colorado River. The LCR MSCP is a partnership of state, federal, Indian Tribal, local, and environmental organizations that represent water, power, and environmental interests along the Colorado River in Arizona, California, and Nevada. The focus of the LCR MSCP is to develop a multi-species conservation plan by December 2000 that will guide conservation activities on the Colorado River for the next 50 years. The cost to develop the Plan will not exceed $4.5 million, which includes development of the Multi-Species Conservation Plan (MSCP) as well as implementation of interim conservation measures.

Introduction

The question asked by the conference moderator and stated as the title of this paper, "Can a Multi-Species Conservation Plan Conserve 102 Species Including the Four Big River Fish at a Reasonable Cost" is premature. If the question were re-stated to ask, "Will a Multi-Species Conservation Program for the Lower Colorado River work and will the cost be reasonable?", the answers to these two questions would be: 1) we believe that this type of program will work and can serve as a model for addressing endangered species issues in a cooperative process; and 2) it is yet to determine if the costs will be reasonable because the Plan has not been developed. But, I would suggest to you that those parties who ultimately will be paying their portion of the costs associated with implementation of the program would not be involved, if they did not expect the Plan to be implemented at a reasonable cost.

---

[1]Executive Director, Colorado River Board of California, 770 Fairmont Avenue, Suite 100, Glendale, CA 91203

With that in mind, some background on the Lower Colorado River Multi-Species Conservation Program (LCR MSCP) and the expectations that can be achieved are in order.

The Program

In November 1993, in response to an order from the United States District Court, the U.S. Fish and Wildlife Service published a proposed rule to designate critical habitat for four species of native big river fishes listed as endangered in the Colorado River Basin. These four fishes are: the Colorado squawfish, razorback sucker, bonytail chub, and humpback chub. The critical habitat designation included nearly the entire mainstream of the Colorado River and substantial portions of the River's major tributaries. The proposed critical habitat designation was adopted, with minor modifications, in a final rule published by the U.S. Fish and Wildlife Service in March 1994.

Because of the proposed designation of critical habitat for the four species of fishes and the anticipation of additional endangered species issues being raised, the three Lower Colorado River Division States (Arizona, California, and Nevada) and water and hydroelectric power resources users within those states formed a partnership to participate in the development of a long-term comprehensive endangered species management program while accommodating the management and potential future development opportunities of water and hydroelectric resources within the Lower Colorado River Basin. A partnership Memorandum of Understanding (MOU), establishing a steering committee and its goals and objectives was executed in November 1994. The MOU's mutually equal goals included the following:

- Conserve habitat and work toward the recovery of included species within the 100-year flood plain of the lower Colorado River, pursuant to the Endangered Species Act (ESA), and attempt to reduce the likelihood of additional species listings under the ESA; and

- Accommodate current water diversions and hydroelectric power production and optimize opportunities for future water and power development, to the extent consistent with law.

In September 1994, the State parties of Arizona, California, and Nevada commissioned a Feasibility Assessment Study to identify and analyze alternative strategies for the management of endangered species within the ecosystem of the lower Colorado River. This study was completed in December 1994, with a strong recommendation to proceed with the development of a proactive long-term multi-species conservation planning effort.

In August 1995, representatives of Steering Committee executed a

Memorandum of Agreement (MOA) for the development of a Lower Colorado River Multi-Species Conservation Program (LCR MSCP) and formally designated the Steering Committee as the LCR MSCP Steering Committee. The signatories to the MOA were: the Department of the Interior, the Colorado River Board of California, the California Department of Fish and Game, the Arizona Department of Water Resources, the Arizona Game and Fish Department, the Colorado River Commission of Nevada, and the Nevada Division of Wildlife. The MOA provided a three-year commitment of funding for the implementation of interim conservation measures (ICMs) and development of a long-term (50-year) conservation program. The acknowledged purpose of the LCR MSCP was to meet the goals outlined in the original MOU and provide long-term Endangered Species Act (ESA) compliance for federal and nonfederal users of the resources of the lower Colorado River.

In order to address specific concerns of environmental and conservation organizations and gain their support and participation in the LCR MSCP process, the Steering Committee authored a Memorandum of Clarification in June of 1996, which was executed by the seven original signatories of the MOA. The Memorandum of Clarification provides assurance that conservation measures need not be limited to those measures funded by the LCR MSCP and that all activities of the program will be consistent with applicable law.

In June 1996, a Cost-Sharing Agreement was executed by representatives of the Department of the Interior and the three Lower Division States to formalize a federal-nonfederal cost-sharing effort for development of the LCR MSCP. The program budget for development of the LCR MSCP was set at $4.5 million over the next three to five years and includes implementation of interim conservation measures (ICMs) targeted specifically at on-the-ground activities and development of the long-term Multi-Species Conservation Plan (MSCP). The Cost-Sharing Agreement provides for a 50-50 sharing of funding between the federal and nonfederal partners. For Year-1, the State entities are providing the nonfederal share through voluntary contributions by the water and hydroelectric power users with interest in the lower Colorado River.

The LCR MSCP's planning area, as defined in the MOA, comprises the mainstream of the lower Colorado River below the Glen Canyon Dam to the Southerly International Boundary with the Republic of Mexico and includes the 100-year flood plain and the full-pool water surface elevation of each mainstream reservoir.

The planning area and the participating interest groups represent a diverse array of land ownership and management authorities and practices. These entities range from Federal interests under the jurisdiction of the National Park Service (NPS), the U.S. Fish and Wildlife Service (USFWS), the Bureau of Land Management (BLM), and the Bureau of Reclamation (USBR), to interests and jurisdictions of Indian Tribes, State water and wildlife agencies, water and power contractors in each of the three states, and environmental and conservation groups. Recreational users of the lower Colorado River (sportfishing, boating and rafting, etc.), private landowners, and local and elected officials are expected to become involved in the process.

The LCR MSCP Steering Committee, through the MOA, adopted a list of 102 species that currently occupy or have historically occupied the planning area for inclusion in the overall program. The list of species includes Federal and State listed sensitive, threatened, and endangered species.

In January 1997, the USFWS designated the LCR MSCP Steering Committee as an Ecosystem Conservation and Recovery Implementation Team (ECRIT). The designation is pursuant to section 4(f)(2) of the ESA which authorizes the Secretary of the Interior to procure the services of appropriate public and private agencies, institutions and other qualified persons to implement recovery actions. The USFWS, in conveying the ECRIT designation to the LCR MSCP Steering Committee, recognized that this effort is unique and encompasses a diverse group of stakeholders who support ecosystem-based conservation of the lower Colorado Rivers' resources. Additionally, designation as an ECRIT provides a lawful exemption, under the ESA, to the requirements of the Federal Advisory Committee Act. Development and implementation of the long-term MSCP by the Steering Committee has the potential to meet the needs of the human residents as well as the needs of fish, wildlife, and associated habitat throughout the Lower Colorado River Basin.

Over the past four years the planning process has progressed steadily. The partnership is established; funding is in place to develop the MSCP for the long-term program; the non-profit National Fish and Wildlife Foundation (NFWF) has agreed to serve as the fund manager and contract administrator; interim conservation measures to provide immediate benefit to species are being implemented; the LCR MSCP Steering Committee has been designated as an ECRIT; and contracts were awarded in September 1997 to develop the fifty-year MSCP and hire a program facilitator.

Now that the program is in place and the plan is being developed, what are the expectations? The MSCP is expected to be completed by December 2000. The costs for developing this Plan and providing ICMs for the next three years will not exceed \$4.5 million. This is clearly a reasonable cost to develop the MSCP. The Federal participants will provide 50% of the program development costs or \$2.25 million. The remaining \$2.25 million will be shared among the nonfederal participants: 50% by California, 30% by Arizona, and 20% by Nevada. Both federal and nonfederal Year-1 funding, in the amount of about \$1.5 million, was deposited with the National Fish and Wildlife Foundation in June 1997.

- Arizona's total share of MSCP development costs is $675,000, or 30% of the nonfederal share of $2.25 million. Arizona's Year-1 Program development costs were voluntarily borne by agencies with interest in Colorado River hydroelectric power generation benefits and water deliveries. Contributors currently include the Arizona Power Authority, Central Arizona Water Conservation District, Salt River Project, and Yuma-area interests.

- California's total share of the program development costs is $1,125,000. California's costs have been voluntarily provided by water and power agencies within southern California, including the Metropolitan Water District of Southern California, Los Angeles Department of Water and Power, San Diego County Water Authority, Palo Verde Irrigation District, Imperial Irrigation District, Coachella Valley Water District, Southern California Public Power Authority, Southern California Edison Company, and San Diego Gas and Electric Company.

- Nevada's total share is $450,000 and includes contributions from the Southern Nevada Water Authority, Nevada Power, Timet, and Kerr McGee Company.

Implementation of the MSCP is expected to provide the biological assurances that the endangered species being considered in the lower Colorado River will progress towards recovery. In addition, sensitive species will be sustained at a level that precludes the necessity for additional listing under the ESA.

The plan is expected to provide assurance to the water and hydroelectric power agencies that current operations can continue and that future uses of the resources provided by the River will be available to them.

The MSCP is expected to guide and coordinate the various fish, wildlife, and habitat recovery or enhancement efforts on the lower Colorado River for the next fifty years. As noted earlier, development of the Plan will not exceed $4.5 million. Clearly this appears to be a reasonable cost and the participating entities expect to keep program implementation costs reasonable. However, the final determination regarding the reasonableness of the costs must wait until the MSCP has been developed, three years hence.

<u>Conclusion</u>

In summary, the Lower Colorado River Multi-Species Conservation Program is extremely important to all of the inhabitants of the desert southwest. It is an innovative and proactive program intended to address endangered species and habitat needs, prevent the need for future listings of species under the ESA, and provide water and hydroelectric power users their current as well as future uses of the resources provided by the Colorado River. The immediate and long-term benefits of a successful MSCP will be felt by the species and the ecosystem of the lower Colorado River, the agricultural interests in southern California and Arizona, the metropolitan regions of Los Angeles, San Diego, Phoenix, Tucson, and Las Vegas, as well as, others who enjoy the resources provided by the Colorado River.

Coordination of the CALFED Bay-Delta Program

Steve Yaeger[1] and Ron Ott[2]

Abstract

The CALFED Bay-Delta Program is a coordinated effort between planners and environmental regulators to develop a solution to environmental and water management problems of one of California's most precious resources, the San Francisco Bay, Sacramento-San Joaquin River Estuary (Bay-Delta system). It is an historic, collaborative environmental planning effort bringing together California interest groups and the state and federal governments to resolve the water supply, water quality, ecological health, and levee system integrity problems associated with the Bay-Delta system. A consortium of 16 state and federal agencies with management and regulatory responsibilities in the Bay-Delta has come together in a collaborative effort with Bay-Delta stakeholders, urban and agriculture water users, fishing interests, environmental organizations, businesses, and other public interests to develop a long-term Bay-Delta solution. With the diverse agencies providing staff and financial resources, a massive coordination effort was implemented to insure that the program could reach a solution in a historic time frame. This paper discusses the elements of the coordination program required to develop a long-term solution to Bay-Delta ecological and water management problems.

Introduction

The Bay-Delta is the largest estuary on the West Coast ,a beautiful, lush , and varied ecosystem including a maze of 1,100 miles of tributaries, sloughs, and islands encompassing 738,000 acres. Lying at the confluence of California's two largest rivers the

----

[1]Steve Yaeger P.E.,Deputy Director, CALFED Bay-Delta Program, 1416 Ninth Street, Suite 1155, Sacramento, CA 95814
[2]Ron Ott Ph.D.,P.E., CH2M HILL, Consultant Team Manager, CALFED Bay-Delta Program, Sacramento, CA 95814

Sacramento and the San Joaquin, it is a haven for plants and wildlife, including 70,000 acres of wetlands and supporting 120 fish and wildlife species. The Bay-Delta is critical to California's economy, supplying drinking water for 23 million California's and irrigation water for over 4 million acres of the world's most productive farmland, producing 45 percent of the nation's produce.

For decades, this area has been the focus of competing interests -- economic and ecological, urban and agriculture. As a result, habitats are declining, and several native species are endangered. The system no longer serves as a reliable source of high-quality water, and the levees that protect the water supply and surrounding agriculture, habitat, and infrastructure face an unacceptable high risk of breaching. Though many efforts by federal and state agencies and interest groups have been made to address these problems, the issues are complex and interrelated, and many remain unresolved.

Program Organization

The CALFED Bay-Delta Program was established in May 1995 and is one element of CALFED, a consortium of state and federal agencies with management and regulatory responsibilities in the Bay-Delta. Member agencies are shown below:

CALFED Agencies

State

Resources Agency
- -Department of Water Resources
- -Department of Fish and Game

Environmental Protection Agency
- -State Water Resources Control Board

Federal

Department of the Interior
- -Fish and Wildlife Service
- -Bureau of Reclamation
- -Bureau of Land Management

Department of Agriculture
- -Natural Resources Conservation Service

Environmental Protection Agency

U.S. Army Corps of Engineers

Department of Commerce
- -National Marine Fisheries Service

U.S. Forest Service

CALFED provides policy direction to the Program. It was formed as part of a Framework Agreement sign in June 1994 by the California Governor and the Secretary of the U.S. Department of the Interior. As part of this agreement, the state and federal governments pledge to work together to formulate water quality standards to protect the Bay-Delta, coordinate the state and federal water projects operations in the Bay-Delta, and develop a long-term Bay-Delta Program.

In December 1994, the Bay-Delta Accord was signed by state and federal regulatory agencies, with the cooperation of diverse interest groups, to address these issues. The CALFED Bay-Delta Program was formed and charged with the responsibility of the later issue: development a long-term Bay-Delta solution.

Coordination Structure

With 16 diverse agencies providing staff and financial resources, a massive coordination effort was implemented to insure that the program could reach a long-term solution in a historic time frame. This Program coordination involves four general areas: policy, technical, stakeholder participation, and public information. Each of the cooperating agencies has management and regulatory responsibilities in the Delta and therefore participate in all four areas. They also participated with the stakeholders to set a clear Program mission statement, goals, objectives, and solution principles. Coordination was essential between the Program agencies and stakeholders in using multiple sources of funding for early implantation projects and the long term solutions.

Policy for the Program is set by the Policy Group. The group consists of the heads of the participating agencies and is co-chaired by California Secretary of Resources and the U.S. environmental Protection Agency Assistant Administrator for Water. A subset of the Policy Group, the Management team, pre-reviews issues and frames them for consideration of the policy makers. Recommendations from the Policy Group go directly to the Governor's and Secretary of the Interior's offices.

As shown in the Program Structure chart below, input to the Management team comes from the Program staff and the Bay-Delta Advisory Council (BDAC). Public input to the program primarily comes through BDAC which includes representatives of some of the most important Bay-Delta stakeholder groups. Smaller work groups in BDAC make recommendations to BDAC on such subjects as finance, water use efficiency, ecosystem restoration, and assurances, which assure that the program will be implemented and operated as planned.

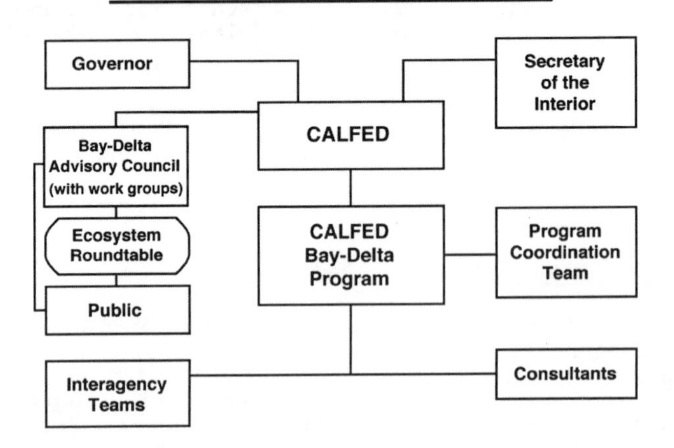

The public is also involved in extensive workshops on specific subject areas, and in general outreach and scoping programs. Stakeholders and agencies also make up the Ecosystem Roundtable that makes recommendations on priories and implementation of projects for the ecosystem.

Technical coordination between the Program agencies and staff, and stakeholder experts is accomplished through technical workshops, technical teams, the Program Coordination Team (PCT) and the Bay-Delta Modeling Forum.

Technical experts from each of the member agencies, made up of PCT. It reviews and makes recommendations on technical information presented by the Program staff. Members of the PCT are also charged with involving the staff in their agencies and to keep the agency heads apprized of the technical aspects of the Program. The modeling forum is an independent group of computer modelers that review the technical aspects of the computer models used in the Bay-Delta system. Other technical input is provided by interagency technical teams and consultants. These teams addressed subject areas such as conveyance/storage, ecosystem, water quality, levee and channels, permit streamlining, and large diversion fish screens.

Program Coordination

The key to successful coordination between the multiple members of the program was the early establishment of the Program's mission, goals, objectives, and solution principles. These were established in public workshops, improved in BDAC, and approved by the Policy Group. The mission of the CALFED Bay-Delta Program is

> *to develop a long term comprehensive plan that will restore ecological health and improve water management for the beneficial uses of the Bay-Delta system.*

Four primary areas of concern that the Program is addressing are **Water Quality**, **Ecosystem Quality**, **Water Supply Reliability**, and **System Vulnerability**. The corresponding primary goals to address these concerns are

- to provide good water quality for all beneficial uses;
- to improve and increase aquatic and terrestrial habitats and improve ecological functions in the Bay-Delta to support sustainable populations of diverse and valuable plant and animal species;
- to reduce the mismatch between Bay-Delta water supplies and current and projected beneficial uses dependent on the Bay-Delta system;
- to reduce the risk to land use and associated economic activities, water supply, infrastructure, and the ecosystem from catastrophic breaching of Delta levees.

These goals were further refined into objectives and sub-objectives which help define the success of the Program. Each alternative solution is being evaluated for it's ability to meet these objectives.

While the objectives are technical in nature, the Program developed six solution principles that offer broad policy guidance. These serve as the criteria for any Bay-Delta solution. According to the solution principles, a solution must:

- *Reduce Conflicts in the System*
- *Be Equitable*

- *Be Affordable*
- *Be Durable*
- *Be Implementable*
- *Have No Significant Redirected Impacts*

Decision process

Once the Program had defined the conflict areas and determined the goals and criteria for a long term Bay-Delta solution, actions were developed to met the goals. These actions, to the greatest extent practicable, would provide multiple benefits to all resource areas. Actions were then combined into numerous alternatives that were then reduced to three comprehensive alternative solutions to the Bay-Delta problems. The three general solutions were then expanded into 17 detail alternatives variations that represented the full range of possible solutions.

To reduce this range to a single alternative, a decision process was developed and applied to all alternatives. The process included determining the performance of each alternative to meet the program objects and solution principles.

Conclusion

Given the complexity of the Bay-Delta and the multiple agencies and stakeholders involved, there were many coordination issues. The time factor was an additional hindrance to coordination. Because of limited time, many sequential aspects of the Program had to be accomplished in parallel. For example, computer modeling being completed at the same time as the impact analysis was being written required intensive coordination between the agency and stakeholder modelers and impact staff.

The agencies involved are large with multiple branches and functions. With the fast pace of the Program, many times the information from the technical member of the agency would not be clearly transmitted to the policy maker in the agency. This resulted in many decisions on issues that had to be recycled through the structure until every one was on board with the decision.

A well defined structure, with clear Program mission, goals, and objectives, along with the guidance of the solution principles, were key to the coordination effort. Even with the multiple coordination issues, the CALFED Bay-Delta Program successfully selected a programmatic alternative to carried forward into detailed analysis and refinement.

Optimization Modeling for River Restoration

C.L. Lowney[1], J.R. Lund[2], M.L. Deas[1]

Abstract

Heightened environmental concerns and regulation in recent decades has led to increased efforts towards environmental rehabilitation of river systems. In pursuing these relatively new environmental objectives, we often find our long experience with managing water for traditional uses to be an imperfect guide. Thus, we require a means of bringing in those more expert with these new operating objectives (e.g., biologists, ecologists) together with analytical techniques that allow us to better understand how our water resource systems can operate for meeting society's expectations.

The purpose of this paper is to examine the utility of systems analysis techniques of simulation and optimization modeling for environmental restoration of river systems with minimal reduction in existing traditional water uses. A linear programming (optimization) model is developed for the Battle Creek system, a tributary to the Sacramento River in California, to identify efficient trade-offs between salmon and hydropower production during the month of August. Currently, salmon suitability is measured in terms of discharge requirements for spawning and summering habitat. For this run-of-river system, with negligible storage, a series of static linear programs can solve this problem within a spreadsheet. Important additional concerns, not included in this screening model, are water temperature, salmon lifestages and the removal of hydropower facilities.

---

[1] Doctoral Candidate, Department of Civil and Environmental Engineering, University of California, Davis, CA 95616

[2] Professor, Department of Civil and Environmental Engineering, University of California, Davis, CA 95616

Background

Prior to the 1940's, tributaries in the upper Sacramento River basin drained the volcanic region of Mount Shasta, and were large and cold enough to support salmon throughout the summer months. Construction of Shasta Dam blocked this upstream habitat forcing these fish to spawn in downstream tributaries. Battle Creek is situated in a volcanic region, and as such is uniquely similar to the habitat now blocked by Shasta Dam, characterized by the numerous large spring flows which sustain substantial baseflow throughout the summer and into early fall. In contrast, other remaining tributaries downstream of Shasta Dam are alluvial streams with substantially lower base-flows and higher water temperatures during the summer months.

Due to its significant gradient and sustained base-flow, Battle Creek is well suited for run-of-river hydropower, which began in 1901 with construction of Volta I, Volta II, South, Inskip and Coleman Powerhouses. The present Battle Creek project consists of these five powerhouses as well as two small storage reservoirs (North Battle Creek and McCumber), three forebays (Grace, Nora and Coleman) three large diversions on North Fork Battle Creek (North Battle Creek Feeder, Wildcat Canal, and Eagle Canyon Canal), three diversions on South Fork Battle Creek (South, Inskip, and Coleman Canals), and numerous tributary diversions, shown below in Figure 1.

Re-Operation Objectives for Battle Creek

Anadromous fish production in Battle Creek is in part limited by stream flow. Bypass flows required by the Federal Energy Regulatory Commission (FERC) project license are 3 cubic-feet-per-second (cfs) in the North Fork, and 5 cfs in the South Fork. Increased flows of 30 to 50 cfs are required for salmon to spawn and rear successfully (USFWS, 1995). An optimization model, which uses the spreadsheet's linear solver, finds the hydropower maximizing distribution of available water to the five powerhouses, while still satisfying instream flow requirements can be used to assist in the evaluation of possible re-design scenarios.

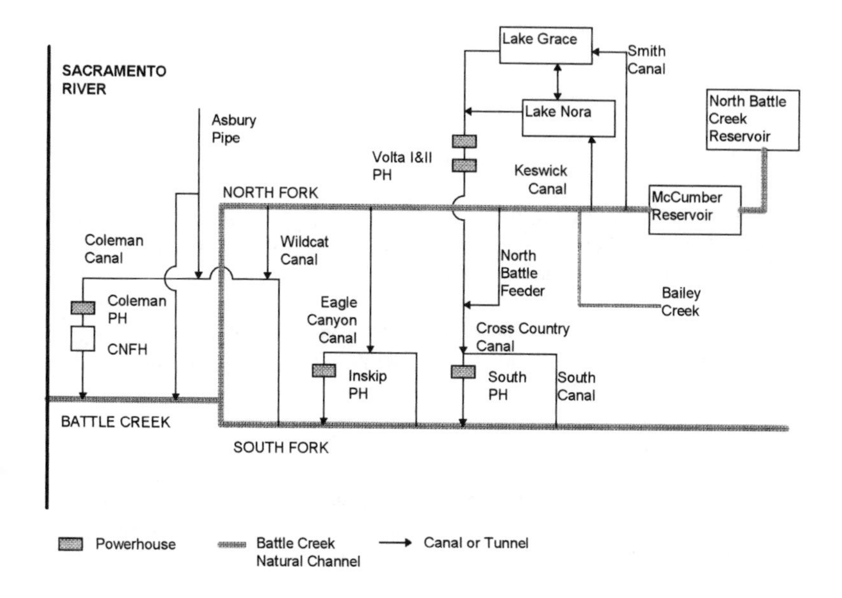

Figure 1 - Schematic Diagram of Battle Creek, Canals and Powerhouses

Model Description

The Battle Creek system consists of the main stem, North and South Forks, canals, penstocks, tunnels and reservoirs. Two small reservoirs, McCumber and North Battle are present on the upper North Fork, however, their combined storage is only 1.5 TAF, thus the system is considered run-of-river for this analysis and may be reduced to canals and reaches (as shown in Figure 1), with canal flows as decision variables, and instream flow requirements in each reach as constraints. Power revenues are maximized by distributing available flow through the system while satisfying instream flow requirements. For the specific case of five powerhouses, the objective function may be written as,

$$\text{Max}\sum_{i=1}^{5}\left(\frac{e_i Q_i h_i}{k}\right)$$

where i is each power plant, e_i is the turbine efficiency, Q_i is flow in the penstocks, h_i is potential head, and k is a unit conversion constant.

Constraints on the system are

$C_j \leq CAP_j, \forall j$

$R_n \geq IFR_n, \forall n$

and mass balance constraints at each junction,

where j is the number of canals, C_j is the flow in each canal, CAP_j is the capacity of each canal, R_n is the flow in any reach, IFR_n is the instream flow requirement for each reach and n is the total number of reaches.

As a static model, with no storage operations, the model is run for a particular constant inflow hydrology. Typically, a percentile frequency of a particular months inflows are used as input hydrology.

Applications

As an example of the application of the linear program, median August flows were used as input hydrology for all model runs (Payne, 1991). Therefore, all applications illustrate system behavior during the dry season only.

Using the linear program, maximum hydropower revenue was computed for various instream flow requirements. Because instream flows are met by decreasing flow diverted through power canals, power revenue is reduced as instream flows increase. Results are expressed as a percentage reduction in power revenue, using current instream flows (3 cfs in the North Fork, 5 cfs in the South Fork) , as the baseline for percentage calculations.

Figure 2 illustrates the potential percentage reduction in power revenue that would result when instream flows are increased beyond baseline conditions. The two curves shown on this plot represent reduction in revenue that would occur if instream flows were increased in both the North and South Fork, and the reduction in revenue that would occur if instream flows were increased in the North Fork only.

Figure 3 shows the percent reduction in power revenue when flows in the North Fork are increased incrementally, working upstream from the confluence of the North and South Forks. The three curves illustrate the percentage reduction in revenue resulting from an increase in instream flow from 3 cfs to 10 cfs, to 20 cfs and to 25 cfs, respectively.

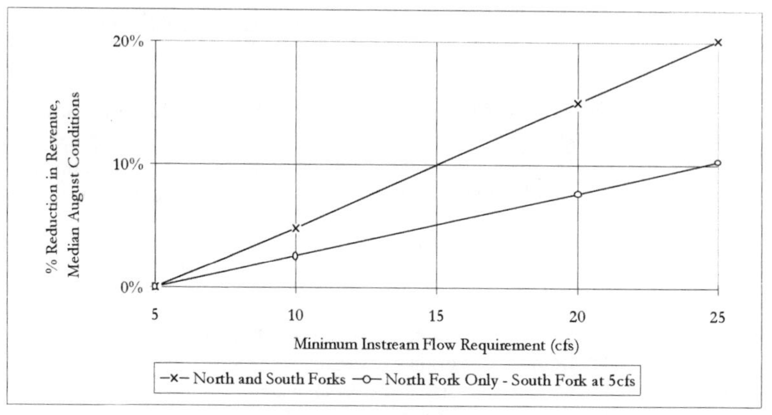

Figure 2 - Estimated reduction in revenue for various flow scenarios, median August inflows

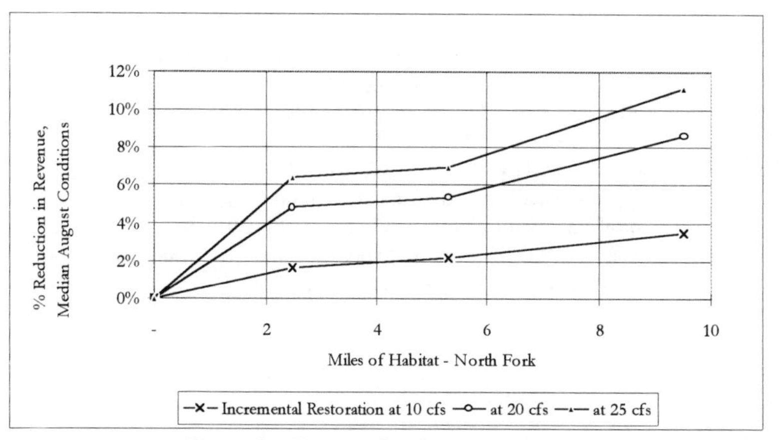

Figure 3 - Estimated reduction in revenues for incremental restoration scenario, median August inflows

Discussion

This model describes the Battle Creek system under median August flow conditions only, and as such does not represent reduction in power revenue during an entire year. The model could be expanded to evaluate alternatives over all months and ranges of flow frequencies. Multiple years and months are easily incorporated through consecutive model runs, automated using spreadsheet macros.

Although power generation was the single objective in this preliminary model, additional considerations could be included, so long as they are easily incorporated into a spread-sheet application. A water temperature model could be included to ensure that target temperatures are met for various flow conditions.

Conclusions

This spreadsheet model illustrates in a small way how optimization models can be used to help design and evaluate stream restoration alternatives. In addition to providing a tool to assist in decision making, modeling forces the explicit examination of system interactions and management objectives. Optimization models are useful for investigating operation for new system management objectives, such as the incorporation of restoration requirements.

References

Thomas R. Payne & Associates. 1991. DRAFT Hydrologic Analysis for Battle Creek, Shasta and Tehama Counties. March 22, 1991. Prepared for California Department of Fish and Game. Redding, CA.

U.S. Fish and Wildlife Service. 1995. Working Paper on restoration needs: habitat restoration actions to double natural production of anadromous fish in the Central Valley of California. Volume 3. May 9, 1995. Prepared for the U.S. Fish and Wildlife Services under the direction of the Anadromous Fish Restoration Program Core Group. Stockton, CA.

POTOMAC

Moderator, Michael A. Ports

"Paradigm Shift in Stormwater Management Regulations"
Daniel J. O'Leary
Parsons Brinckerhoff

"Tributary Strategies in the Chesapeake Basin"
Michael A. Ports
Parsons Brinckerhoff

"Identifying Cost-Effective Watershed-Scale Nutrient Reduction Strategies for the Potomac River Basin"
Stuart S. Schwartz, Ph.D.
Hydrologic Research Center, San Diego

"Public Involvement in Chesapeake Bay Restoration"
Glenn Page and Ryan C. Davis
Alliance for the Chesapeake Bay

PARADIGM SHIFT IN MARYLAND STORMWATER MANAGEMENT REGULATIONS

Daniel J. O'Leary[1], P.E., M.ASCE, Brian S. Clevenger[2]

ABSTRACT: The State of Maryland adopted a Stormwater Management Law[1] in 1982 and accompanying Regulations[2] in 1983 to control the effects of runoff increases caused by new development. After 14 years of a formalized stormwater management program, the Maryland Department of the Environment (MDE) is in the process of updating not only the State's current Regulations but shifting on how urban runoff is controlled. Reasons for a major philosophical change in Maryland's stormwater program include accommodating changes in technology, phasing out practices that have not performed as anticipated, addressing more strict Federal National Pollutant Discharge Elimination System (NPDES) requirements, providing a uniform foundation for local programs statewide, and injecting over 14 years worth of implementation experience. This paper includes a brief history of stormwater regulation in Maryland, some lessons learned, as well as what regulation changes can be expected in Maryland, and possibly other states, in the not too distant future.

History

Maryland's current stormwater management program was developed as part of an overall effort to protect and help clean up the Chesapeake Bay. This resulted in passage of a formal Stormwater Management Law[1] in 1982. Another, separate Maryland law aimed at controlling sediment runoff from construction sites had already been in effect since 1970[3]. In 1982, the prevailing attitude was that if flooding caused by increases in runoff volume from new development (and sediment runoff from construction sites) were controlled, then the quality of receiving streams could be maintained. As a result, the original (and current) Regulations adopted into the Code of Maryland Regulations require only that runoff quantity be controlled. Quantitative control is to be achieved through the use of certain best management practices (BMPs) that inherently include water quality benefits. Minimum design criteria for acceptable BMPs is presented in the Regulations. The Maryland Department of Environment (MDE), Water

---

[1] Senior Water Resources Engineer, Parsons Brinckerhoff, 301 N. Charles Street, Suite 200, Baltimore, Maryland 21201, (410) 385-4178, fax (410) 727-4608, e-mail: oleary@pbworld.com
[2] Environmental Program Administrator, Maryland Department of the Environment, 2500 Broening Highway, Baltimore, Maryland, 21224, (410) 631-3543, fax (410) 631-3553

Management Administration is responsible for administering the statewide stormwater management program.

Maryland's Current (1997) Policy

Quantity stormwater management requirements established Maryland's Regulations focus on the 2-year and 10-year storms. Special consideration is given to the State's Eastern Shore (east of the Chesapeake Bay) and to "interjurisdictional flood hazard watersheds," where historical flooding damages have been documented. The current policy applies to all development projects that disturb over 5,000 square feet of land area, except exempted activities as noted below. For the majority of Maryland, post-development 2- and 10-year storm peak discharges must be "maintained at a level equal to or less than the respective 2-year and 10-year pre-development peak discharge rates." Most Eastern Shore counties are required to address only the 2-year storm and developments within interjurisdictional flood hazard watersheds must address 2-year, 10-year, and 100-year return period storms.

Acceptable BMPs, in order of preference, listed in the Regulations include infiltration of stormwater as most preferred, flow attenuation by use of open vegetated swales and natural depressions second, stormwater retention structures third, and stormwater detention structures as least preferred. The Regulations require that developers first investigate the use of the most preferred BMP (infiltration) on all sites in accordance with feasibility criteria outlined in the Standards and Specifications for Infiltration Practices[4] developed by MDE. If infiltration of stormwater was determined not to be feasible due to unfavorable site characteristics, then the developer would move down the list of acceptable BMPs until one that was appropriate for the particular site was reached. Documentation supporting the decision to use lower preference BMPs is required to be submitted with stormwater management plans for approval.

Included in the Regulations are provisions for exemptions, waivers, and variances. Exemptions allow certain types of land disturbing activities to occur without adherence to stormwater management requirements (e.g., disturbances of less than 5000 square feet, certain agricultural activities, etc.). Waivers lessen stormwater management requirements depending on development circumstances. Waivers on a case by case basis may be granted for projects demonstrating no adverse impacts to receiving streams. Waiver decisions are made by the approval authority having jurisdiction over the project. Typically, these decisions are rendered by local governments who have direct responsibility for administering Maryland's stormwater management program through plans review and approval and construction and maintenance inspections. MDE reviews, approves, and inspects plans for State and Federal projects and evaluates the effectiveness of county and municipal stormwater management programs.

Variances allow stormwater management requirements to be reduced or eliminated for a given site if significant hardship renders stormwater management impractical or if the intent of the Regulations can not be met. Variances are granted on a case by case basis. "Fees in lieu", which allow developers to pay fees in place of including stormwater management on site, are available in some localities.

Because the Regulations were originally water quantity control oriented, there were no specific provisions for protecting the quality of water leaving development sites. While new development was required to control stormwater volume increases, redevelopment sites routinely qualified for waivers because significant enough increases in impervious surfaces to post-development discharges were rare. This led the State to amend the Stormwater Management Law in 1988 to allow water quality improvement practices to be required "even when existing runoff characteristics are maintained."

Stormwater management policy has, over time, departed slightly from the strict letter but not the intent of the Regulations through creative interpretations to address programmatic shortcomings and unanticipated development scenarios. For example, the use of "flow attenuation by use of open vegetated swales and natural depressions" to control stormwater is not easily quantified. Therefore, at the State level this is not recognized as an allowable practice. Additionally, detention of stormwater has very little water quality benefit when the 2-year and 10-year storms are being analyzed, so "detention" has been replaced by some localities and the State with "extended detention" in accordance with MDE criteria[5].

Why Change?

By the late 1980's, stormwater quality issues were becoming more prominent. Strategies to control nutrient inputs into the Chesapeake Bay were refined to include ten major tributaries, some of which were suffering from stormwater related problems; the Federal NPDES program was being modified to include stormwater as a point source discharge; Maryland's 401 Water Quality Certification Program was under attack by critics for not demanding higher standards where stormwater discharges were directed toward wetlands; and, after 10 years of formal stormwater management regulation, developers were controlling 2-year and 10-year return period storm peak discharges at great expense while being blamed for degradation in receiving streams. Additionally, with each county regulating stormwater individually, stormwater management resources were being stretched while a consistent approach was trying to be maintained statewide. Couple these factors with increases in development and water quality can be seriously affected.

What's New?

In 1994 MDE formed the Stormwater Management Regulations Committee (Committee) to review Maryland's existing program and recommend revisions that would address several issues. These include water quality, minimum control criteria, redevelopment, liberal waiver policies, and the lack of standardized design criteria for stormwater management structures statewide. The Committee is comprised of State and county stormwater management regulators, environmentalists, developers, developer associations, and consultant engineers. It has been meeting on a regular basis with subcommittees established to address recognized or perceived weaknesses in the Regulations.

Proposed Changes to Regulations

MDE had a good idea about needed regulatory improvements when the Committee was convened. The Committee's meetings have been very organized by agenda and specific time limits on discussions. The majority of decisions have been made by consensus. Specific areas of the Regulations that have been addressed are as follows:

Water Quality and Design Standards

One means for addressing water quality and stormwater management design standards would be to create a manual, similar to what exists in Maryland for Erosion and Sediment Control[6], for the purpose of complimenting the Regulations. As a group, the Committee agreed with MDE's suggestion that the development of a statewide design manual that could be incorporated by reference into the Regulations could solve a number of the problems currently facing the State's program. Additionally, a design manual would theoretically be easier to modify in the future. Therefore, MDE applied for and received a grant from the Coastal Zone Management Act Section 309 to develop a Maryland Stormwater Design Manual (SW Manual).

Incorporation of the SW Manual into the Regulations by reference would eliminate technical jargon now present in the Regulations and spell out specific water quality and other treatment practice requirements. The SW Manual will also standardize the approach to stormwater management statewide and recognize differences in physiographic regions of Maryland. The Committee is currently investigating minimum control criteria, waiver policies and redevelopment issues for which the SW Manual will provide the technical tools for designing stormwater structures.

In 1995 MDE issued a Request for Proposals to create a stormwater design manual. The firm selected to perform the work is the Center for Watershed Protection, Silver Spring, Maryland. In the fall of 1997 the draft Maryland Stormwater Design Manual was distributed to the Committee members for review and comment.

The SW Manual is divided into Volumes I and II that include stormwater design criteria and technical appendices, respectively. The stormwater design criteria contains methods to calculate stormwater requirements for water quality, groundwater recharge, stream channel protection, overbank flood control, and extreme flood volumes. Also included are minimum design criteria for BMPs, criteria for selecting BMP locations, and environmental site design credits. The technical appendices include landscaping guidance for stormwater BMPs, construction specifications, step-by-step design examples, and assorted design tools to support the stormwater design criteria in Volume I.

Minimum Control Criteria

As mentioned previously, the current stormwater management requirements for the majority of Maryland involve "managing" the post-development 2-year and 10-year storm peak discharges back to existing conditions. Arguably, this policy has done little to protect stream channel erosion because it actually decreases the return period for bankfull storm events[7], assuming bankfull conditions occur more frequently than once every two years. MDE has conceded that the current Regulations do not adequately address streambank erosion so the minimum control criteria in the Regulations should be changed.

The change to the minimum control criteria suggested is to eliminate "managing" the 2-year post-development peak discharge to pre-development rates. This will be replaced with extended detention (ED) which involves designing a basin that separates the center of mass of the post-development 1-year storm inflow and outflow hydrographs by 24 hours[5]. Modeling of the effects of ED by this definition are currently underway to determine the effects of basins designed by this criteria on a wide range of storm events.

Waiver Policy

A subcommittee of the Committee was established to deal with fixing alleged and potential abuses within current waiver policies. The current Regulations state that county ordinances may include waiver provisions that apply to developments on a case by case basis, only if the waiver provisions are approved by MDE. This topic caused some heated debate between developers and regulators and environmentalists. Developers think that waivers are necessary in some cases, regulators like the flexibility that waiver provisions provide, and environmentalists want to abolish them to make every developer responsible for quantity and quality of runoff from their developments. Because the subcommittee on waivers was so diverse, in the end the participants agreed to disagree and leave the development of a waivers policy up to MDE, subject to a vote by the full Committee.

Conclusion

Maryland emerged as one of the first states to formally adopt stormwater management requirements for development. Experience and a desire to improve water quality in receiving streams has driven proposed changes in stormwater management regulation in Maryland. The proposed changes are anticipated to address water quality, minimum control criteria, and waiver policies. A design manual is being developed that is expected to replace wordy technical regulatory language by "incorporation by reference". The SW Manual will simplify the Regulations while providing methods for achieving water quality goals and BMP design guidance and promote statewide stormwater management approach uniformity. Minimum control requirements are expected to change from the current 2-year and 10-year detention of post development discharge peaks to pre-development levels to 1-year extended detention of the post-development discharge to reduce receiving stream degradation. Waiver policies allowed under current Regulations are expected to change but the magnitude and direction of change are still unknown. Redevelopment issues are expected to change in the Regulations to be similar to those for new development, unless existing impervious area is significantly reduced.

Summary

The State of Maryland is proposing revisions to its current stormwater management policies and Regulations that represent a significant departure from the original model. Committee involvement, identification of weaknesses in current Regulations, proposals for strengthening water quality provisions, and the development of a design manual are all steps that have been taken to better address some of the shortfalls of the originally drafted Regulations and resulting problems that have been recognized. Proposed Regulations are anticipated to be available for public comment by the end of 1998.

---

[1] Maryland Environment Article, Title 4, Subtitle 2. Stormwater Management Annotated Code of Maryland.

[2] Code of Maryland Regulations (COMAR) Section 26.09.02.01 through 26.09.02.10.

[3] Maryland Environment Article, Title 4, Subtitle 1. Sediment Control Annotated Code of Maryland.

[4] Standards and Specifications for Infiltration Practices, Maryland Department of Environment, February, 1984

[5] Design of Stormwater Management Extended Detention Structures, Maryland Department of Environment, July, 1987

[6] 1994 Standards and Specifications for Soil Erosion and Sediment Control, Maryland Department of Environment, 1994

[7] McCuen, R.H. (November 1979) Downstream Effects of Stormwater Management Basins. Journal of the Hydraulics Division, American Society of Civil Engineers, Vol. 105, HY-11.

TRIBUTARY STRATEGIES IN THE MIDDLE POTOMAC RIVER BASIN OF THE CHESAPEAKE BAY

Michael A. Ports[1], PE, PH, F. ASCE

Introduction

Maryland's Tributary Strategies are a recent addition to the historic Chesapeake Bay Agreement, signed in 1987 to address the problem of excess nutrients and their impacts upon the nation's largest and most productive estuary. Through the Agreement, the Chesapeake Bay Commission, Environmental Protection Agency, and the states of Maryland, Virginia, Pennsylvania, and District of Columbia formally agreed to work together to achieve a 40 percent reduction in nitrogen and phosphorous reaching the Bay by the year 2000. Other water quality, habitat, and stewardship goals were also adopted.

In 1991, the Chesapeake Bay Program conducted a scientific re-evaluation to assess progress toward the 40 percent goal. The study concluded that, although significant progress had been made, more needed to be accomplished toward controlling nutrient pollution. As a result, in 1992 the Executive Council of the Chesapeake Bay Program directed the Bay partners to develop *"tributary strategies"* - watershed-based plans to reduce nitrogen and phosphorous entering the Bay's rivers.

In Maryland, the objective of the tributary strategies is the introduction of a new working relationship between federal, state, and local governments, business, agriculture, and citizens to improve water quality and enhance habitat. Maryland recognized the need to extend the partnership to include those responsible for local land use decisions. To address the different types of pollution problems in different regions of the Chesapeake Watershed, Maryland divided the State into ten tributary basins. Thus, the State developed specific strategies to reach the 40 percent goal within each tributary basin.

---

[1] Principal Professional Associate, Principal Water Resources Engineer, Parsons Brinckerhoff Quade & Douglas, Inc., 301 N. Charles Street, Suite 200, Baltimore, Maryland 21201

Tributary Strategies and Tributary Teams

Appointed by the Governor, with input from local governments and other interested parties, the teams are composed of representatives from state and local agencies, farmers, business, environmental organizations, federal facilities, and citizens. The teams meet regularly, providing local knowledge essential for implementing best management practices and aiding state and local governments target their programs for improved efficiency and participation. The teams

- Ensure that implementation proceeds on schedule.
- Coordinate participation among the parties.
- Promote understanding through public education.

To illustrate how the tributarties strategy works at the local level, this paper describes the activities and accomplishments of the Middle Potomac River Tributary Team during their first year. The team was faced immediately with two challenges, understanding its mission and role and defining its goals and objectives. The team wrestled with a number of initial uncertainties about its responsibilities and how to achieve its goals. The team's recommendations and conclusions that stem from its first year's effort reveal much progress in the process of self-definition.

The team divided into three workgroups: Urban Watershed Management Workgroup, Rural and Agricultural Workgroup, and Wastewater and Point Source Workgroup. Each workgroup determined its own priorities and set its own goals.

Urban Watershed Management Workgroup

The workgroup concluded that the state should develop programmatic goals and measuring sticks which could then be applied usefully to stimulate and track achievements in living resources protection important to the Bay and to local streams. By themselves, arbitrary 40 percent nutrient reduction goals have no technically supportable relationship to the primary protection needs of most small urban streams. Yet protection of living resources dependent upon these streams remains vital to protecting the Bay eco-system.

The arbitrary political boundaries of the Middle Potomac Basin make little technical sense (Figure 1). The absence of an adequate mechanism to foster dialog with their Virginia and District of Columbia counterparts is contrary to the basic premise that all stakeholders in the watershed must

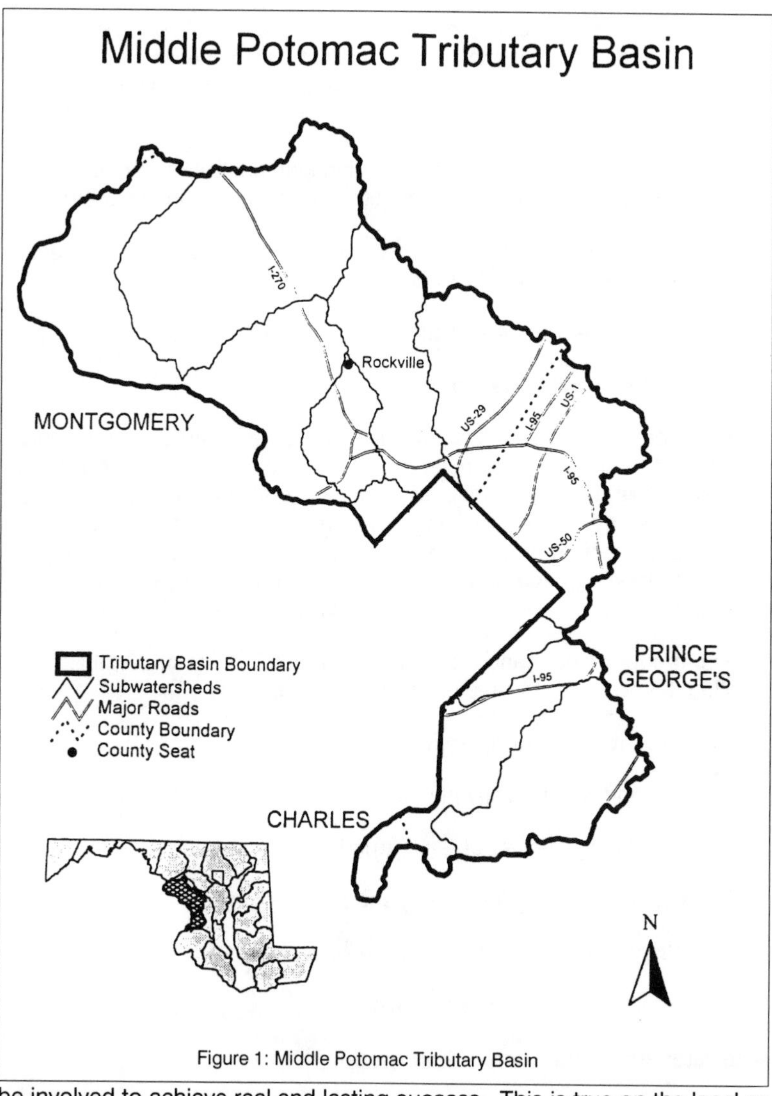

Figure 1: Middle Potomac Tributary Basin

be involved to achieve real and lasting success. This is true on the local as well as the state level.

The workgroup recognized that the NPDES permit program should be linked closely to resource monitoring, watershed inventory, planning, and restoration needs. Various state permitting programs often are not well coordinated and lack a comprehensive watershed protection perspective.

More flexibility is needed to foster trade-offs between permitting requirements that unintentionally work against comprehensive watershed approaches. Further, long-term and reliable funding is needed to support expanded programs for:

- Public education, stormwater management facility maintenance and enforcement, stream monitoring, and pollution enforcement.
- Watershed inventory and feasibility planning to support stream protection and rehabilitation.
- Stormwater retrofit and stream channel restoration.

Rural and Agricultural Workgroup

The workgroup directed its first initiative at horse pasture and manure management. Targeted audiences were identified and written information prepared. Ideas discussed include expanding the cost-share program to include horse operations, linking a horse operations license with appropriate pasture and manure management, planning for a public seminar and best management practice demonstration project. The team co-sponsored a series of seminars with a local Soil Conservation District.

The workgroup recommends that the continued economic viability of agriculture be evaluated. The evaluation should address:

- Incentives and disincentives.
- Soil and water quality plans.
- Inter-family transfer of real property.
- On-farm housing for future generations.
- Alternatives to on-site sewerage systems.
- Transfer of development rights.

Wastewater and Point Source Workgroup

Overall progress in reducing nutrient delivery to the Potomac River is hampered by the uneven pace at which jurisdictions can address nutrient reduction goals. There is no formal mechanism by which Maryland, Virginia, and the District of Columbia can adopt mutually acceptable nutrient reductions with the Basin. Further, point source contributions in the Basin are dominated by those from the Blue Plains Wastewater

Treatment Plant. While the plant treats waste from the three jurisdictions, it is under the jurisdiction of the District government. Continued cooperation among the Blue Plains users, District of Columbia, EPA, should focus on enhanced nutrient removal.

The program has not yet addressed the issue of how nutrient reductions will be maintained, once achieved. The planning horizons for tributary strategies are not the same as those used by local governments. Nutrient trading may become an important mechanism for achieving cost-effective nutrient reductions. Reliable, long-term funding must be continued if progress in meeting nutrient reduction goals is to be made at treatment plants.

Ten Step Strategy

The Tributary Team developed a ten step strategy for reducing pollutant loadings. The strategy requires careful consideration of how all of the pieces fit together. Management efforts that focus on public education, pollution prevention, and protection and restoration of living resources and habitat in small subwatersheds will receive the most public understanding and support and, consequently, the greatest success. The ten steps were presented as a discussion starter. Some of the proposed steps may not be necessary and others may be needed.

1. Divide the watershed into subareas.
2. Characterize by land use or development level.
3. Characterize by biological, chemical, and physical condition.
4. Project changes into the future.
5. Quantify the effects from the projected changes.
6. Rank by resource condition and pollutant potential.
7. Establish priorities for solutions.
8. List potential control strategies.
9. Short list probable effective strategies.
10. Determine implementation process.

Conclusions

The time spent by the team discussing the issues points out that there is no easy solution to the problem. While the problem is complex, the solution is not unmanageable. Part of the solution is legislative, part administrative, and part voluntary. In short, everyone has part to play in the solution. The necessary funding ranges from significant capital investments for wastewater treatment plant upgrades, to more modest investments for stormwater retrofit and stream restoration projects, to entirely volunteer efforts for stream walks and cleanups.

Identifying Cost Effective Watershed-Scale Nutrient Reduction Strategies for the Potomac River Basin

Stuart S. Schwartz Ph.D.[1]

ABSTRACT

Interstate restoration efforts on the Chesapeake Bay have culminated in nutrient reduction goals for each of the Bay's major watersheds. These goals are implemented through the development of tributary-specific nutrient reduction strategies by each of the states participating in the Bay restoration. The Potomac River Basin is unique among the Bay's tributaries in having major portions of its drainage area managed by the States of Pennsylvania, Maryland, Virginia and the District of Columbia. Beyond the coordination of regulatory programs controlling point source dischargers with voluntary cost-share programs for non-point discharges, a Potomac River watershed strategy must also coordinate the competing goals and interests of the respective political jurisdictions.

This presentation describes a comprehensive methodology for identifying cost-effective watershed strategies integrating both point and non-point controls with the institutional and political constraints unique to the Potomac watershed. A consistent cost basis is developed to compare the cost-effectiveness of agricultural and urban non-point management practices with the upgrade of conventional waste water treatment plants to utilize biological and/or chemical nutrient removal. Integrated nutrient control decisions that achieve watershed goals at minimum cost are efficiently identified using multiobjective optimization. When multiple pollutants (e.g. nitrogen, phosphorous, sediment) are considered, coordinated targeting of point source reductions can compliment non-point controls, yielding increased nutrient reductions at lower cost.

Beyond cost minimization, administrative constraints and equity considerations are explicitly considered in Potomac watershed strategies. This is particularly valuable in interstate watersheds where the perception of fairness and burden-sharing can be as important as technical and economic feasibility. In contrast to traditional prescriptive, technology-based approaches, the use of multiple objective optimization supported the identification of interstate strategies that balanced technical and economic efficiency with the equitable targeting of nutrient control measures.

---

[1] Associate Director for Water Resources, Interstate Commission on the Potomac River Basin. **Current Address**: Hydrologic Research Center; 12780 High Bluff Drive, Suite 250; San Diego, CA 92130-2069. (V) 619.794.2726 (FAX) 619.792.2519

Introduction

Growing regional awareness and concern with the widespread degradation of water quality and living resources in Chesapeake Bay culminated in the signing of the Chesapeake Bay agreement. The States of Pennsylvania, Maryland, Virginia, the District of Columbia, and the Federal Government committed to reduce their collective nutrient loads to the Bay to 60% of the 1985 load. Signed by the Governors, the Mayor of the District, and the EPA Administrator, the agreement also called for reevaluation of both the progress and appropriateness of the nutrient reduction goals. Regional commitment to the load reduction targets was reaffirmed in subsequent Chesapeake Bay agreements which extended the nutrient reduction goals to tributary nutrient reduction targets, emphasized the importance of living resources and toxics, and perhaps most importantly set the 40% nutrient reduction as an absolute cap on nutrient loadings, to be maintained beyond 2000.

Watershed strategies: Descriptive Framework

With the formal identification of tributary load targets, emphasis shifted to the description and analysis of basin loadings and nutrient control options. A combination of lumped parameter watershed modeling supplemented by a variety of field scale modeling and monitoring efforts supported the quantitative description of nutrient loading and BMP effectiveness. Recognizing the need to account for cost-effectiveness, a parallel effort developed consistent cost-effectiveness estimates that would allow a fair comparison between alternate nutrient control technologies.

For example, comparing the cost-effectiveness of grassed filter strips to manure storage structures required the capitalization and discounting of both practices on a lifetime common cost basis. In this example, the assumed lifetime of the capital investment in manure storage structures was compared to the present value of the lifetime construction and maintenance cost for a filter strip, which included the "capital" cost of reinstalling a new filter strip every five years (assumed lifetime) over the common 20 year facility life. Through detailed practice-specific analysis, present-value costs of all candidate nutrient control practices for both point source and non-point source loadings were brought to a common discounted lifetime cost basis. This allowed a fair comparison of the relative cost effectiveness of agricultural, and urban best management practices (BMPs) with point source retrofits (upgrading secondary treatment to various levels of biological nutrient removal). Model-based nutrient delivery estimates for each landuse in each Potomac subbasin combined with a consistent common-cost basis for quantifying the cost-effectiveness of every candidate BMP provided the quantitative descriptive framework supporting watershed strategy development.

Watershed Strategies: From Description to Decision

With the consistent quantitative framework describing watershed loading and nutrient control options, attention turned to the challenging task of integrating the technical description of the Potomac watershed management alternatives into specific decisions and allocations of nutrient control technologies to each and every nutrient source throughout the watershed. This process required the constructive participation of resource managers from agricultural non-point, urban non-point, and wastewater discharge management programs from each of States, the District of Columbia, and the Federal Government.

The need for a credible quantitative description of the watershed was clear to all of the stakeholders in the process. The widespread support for a consistent modeling framework to describe and evaluate alternate management decisions is often clearer than the process in which this descriptive tool will be used to reach comprehensive regional decisions. Too often the optimistic assumption of resource managers is that once "***the model***" is completed, the appropriate management decisions will naturally flow from the data. The common experience is that, while the complexity of problems with regional competing interests readily results in support for detailed quantitative models of watershed systems, the quantitative model results demonstrate even greater system complexity in the system, rather than an "obvious" solution. This was certainly the case in the Potomac River basin, in which tradeoffs between agricultural and urban interests, upstream and downstream interests, and regulatory (point source) vs. voluntary (non-point source) management issues were made even more complex by the interstate interests in the shared waters of the Potomac.

Point source dischargers rightly took pride in the significant reductions in nutrient loadings that had already been achieved through a cumulative investment of over $5 B in wastewater treatment, suggesting the need for increased emphasis on non-point runoff in agriculturally rich tributaries such as the Shenandoah and Monocacy Rivers. Non-point dischargers fairly contrasted the large federal subsidies (such as the construction grants program) that had supported upgrades in wastewater treatment plants, to the direct costs imposed on individual farmers by requiring field scale structural controls of non-point runoff. Against this complex background of potentially competing interests, it became clear that, identifying a comprehensive watershed scale nutrient reduction strategy would require a robust decision support framework in addition to the detailed modeling and analysis of basin-wide nutrient loading and BMP cost-effectiveness that had been developed.

Watershed Strategy: Multiobjective Decision Making.

The complex problem of allocating limited pollution abatement resources has most commonly been solved through the implementation of uniform, technology-based standards. The uniform requirements for secondary treatment in publicly owned wastewater treatment plants stemming from the Federal Water Pollution Control Act achieved a major improvement in surface water quality. Technology-based standards have the administrative benefit of being easily described and regulated, and they avoid many -though not all- issues of reasonableness, fairness and equity, by holding all dischargers to a common standard. The cost of this simplification is the foregone opportunity to achieve ambient water quality goals more cost effectively by efficiently targeting limited pollution abatement resources.

A uniform control strategy for the Potomac River basin might, e.g. require all conventionally tilled farmland to employ nutrient management and conservation tillage practices; wastewater treatment plants might similarly be required to uniformly adopt 3-stage biological nutrient reduction. Facing the enormous complexity in allocating nutrient control resources over the Potomac River basin, initial strategies similarly focused on uniform control technologies for all nutrient sources.

The preliminary uniform control strategies for the Potomac successfully achieved the 40% nutrient reduction goal, with estimated *annual* costs of $400 to over $900 million. Faced with this enormous cost, the resource managers were willing to consider strategies in which the allocation of nutrient control alternatives were targeted using mathematical optimization. A variety of "optimal" nutrient control strategies were explored with varying degrees of departure from uniform technology-based standards. For example optimization of non-point BMPs within each State's portion of the Potomac in conjunction with uniform adoption of 3-stage BNR at all point source discharges reduced the estimated annual costs to $244 million.

Optimizing point and non-point source controls within each jurisdiction (while still requiring a 40% nutrient reduction from each State's portion of the Potomac) further reduced the annual cost to $217 million. The absolute least cost solution, in which nutrient control technologies were optimized irrespective of source or political jurisdiction resulted in an annual estimated cost to achieve a 40% nutrient reduction of only $155 million. The dramatic cost savings led to a more serious evaluation of targeted control strategies.

Watershed Strategies: Equity, Efficiency, Implementability.

Closer examination of the least cost solution revealed features that, while technically feasible and economically desirable, were judged to be institutionally infeasible. For example the non-linear scale economies for retrofitting secondary treatment plants sometimes resulted in "optimal" strategies in which two nearby treatment plants of nearly equal capacity required dramatically different levels of treatment. Similarly, comparable farmland in adjoining basins frequently had striking differences in the intensity and cost of BMP application. These conspicuous differences, though economically optimal due to small differences in marginal cost effectiveness, were judged to be politically infeasible due to the perception of inequity in burden sharing. To address these concerns a variety of "equity measures" were formulated and incorporated explicitly in the mathematical optimization problem.

The non-point control problem was modified to equalize the percentage of treated acres, for each landuse (e.g. conventional tilled agriculture, urban, etc.) in each subbasin. For example all conventionally tilled farm land in all sub-basins, in all jurisdictions were required to treat α% of the total acreage. Variations of this constraint were explored that maintained the ability to target the most cost effective BMPs to the acreage treated in each subbasin. In this way a simple, easily described measure of equitable burden sharing was built into the watershed strategy, while exploiting the ability of mathematical optimization to identify the most cost-effective choice of BMPs from the infinite number of possible allocations.

The point source problem was similarly modified by grouping point source discharges into clusters with similar size, cost, and geographic characteristics. For each of the point source "zones" so identified, a single retrofit technology was selected, while allowing the assignment of different technologies to different zones. The inherent differences due to scale economies and differential transport throughout the basin were explicitly considered in choosing treatment levels, while maintaining equitable burden sharing (requiring some level of enhanced nutrient removal from all point sources). These equity constraints addressed many of the pragmatic institutional issues associated with the implementability of a multiobjective multiple-source watershed-scale control strategy. Nevertheless, one significant equity issue remained: interstate equity.

Interstate equity emerged from the recognition that the structural differences in the nature of pollution among the States made it inherently more cost effective for some states to reduce their Potomac River nutrient loads by more than 40%, while allowing other States to achieve a reduction for at least one nutrient of less than 40%. This was understandably viewed as inherently *inequitable*, creating the politically unacceptable perception that one state was paying more than its "fair share". Balanced against this characterization of equity, the complementary nature of point and non-point source nutrient reduction created extremely cost effective nutrient trading opportunities on the basin scale.

Most trading initiatives to date have focussed on single pollutant trades between point and non-point sources, (e.g. phosphorous trading in Dillon Reservoir in Colorado). The complexity of the Potomac River system created enormously richer opportunities to trade among all nutrient sources. While these opportunities were qualitatively recognized, the exploration of meaningful alternatives was only feasible through the use of multiobjective optimization.

By relaxing the constraints requiring a 40% reduction in the nutrient load from each State's portion of the Potomac, the annual cost of achieving the 40% nutrient reduction was shown to drop to within 5% of the absolute minimum cost, while *maintaining* equitable burden sharing among both point and non-point sources, as well as between the states. Moreover, the comprehensive evaluation of all possible nutrient reduction strategies (possible only through the use of multiobjecitve optimization) identified a least cost solution that reduced the costs to *every* jurisdiction. Allowing the *possibility* of inequity in the percentage reduction from each state allowed the identification of complex nutrient trades that reduced the nutrient reduction costs to *all* of the jurisdictions. This was only possible because of the ability of multiobjective optimization to implicitly consider *all possible* resource allocations in a computationally efficient manner.

The use of multiobjecive optimization revealed that the most cost-effective equitable strategy included the effective "trade" of point source nitrogen reductions in the District of Columbia for non-point phosphorous reductions in Pennsylvania. The complex description of the Potomac Basin afforded by watershed modeling and common cost analysis of all nutrient reduction technologies provided the information base that allowed this innovative solution to be identified. However, traditional management approaches based on uniform prescriptive technology-based controls would have excluded this innovative solution from consideration a priori. Without the use of multiobjective optimization, this creative cost-effective alternative would not have been identified. It is equally important to note that if the "credit" of cost-effective nitrogen reduction from the district of Columbia is not applied to Pennsylvania's nitrogen load through this trade, the overall cost of achieving the 40% nutrient reduction increases for *all* of the jurisdictions.

Conclusion

The diversity of watershed-scale control options presents innumerable alternatives for strategy formulation. Increasing marginal pollution removal costs, combined with preferential removal of sediment associated phosphorous in non-point BMPs and dissolved nitrogen in point source retrofits, motivates integrated, comprehensive watershed strategies. The complexity of watershed-scale strategies presents many tradeoffs. This complexity also means many alternatives balancing potentially competing technical, economic, and policy goals may be identified - if tools are available to explore the full range of technical alternatives.

Traditional prescriptive technology-based strategies are familiar and easily described and administered. The simplicity of technology-based approaches derives from the *a priori* exclusion of the vast majority of feasible alternatives, trading high compliance costs for "simplification". Multiobjective optimization can effectively identify the efficient set of feasible strategies without excluding *any* pollution control alternatives a priori. By explicitly including equity and burden sharing in optimal watershed-scale strategies, the objectives of economic efficiency, technical effectiveness and political implementability can be explicitly incorporated in the development of cost effective watershed scale strategies to achieve regional water quality goals.

PUBLIC INVOLVEMENT IN CHESAPEAKE BAY RESTORATION

Glenn G. Page[1]
Ryan C. Davis[2]

Abstract

From 1995-1997, the Alliance for the Chesapeake Bay conducted a three year pilot program to investigate if citizens could assist in the generation of quality assured data to characterize the habitat requirements for Submerged Aquatic Vegetation (SAV) in select tributaries of the Chesapeake Bay. After three years of data collection the data has met quality control standards and has generated quality assured data that has been effectively used to quantify water quality parameters and has also been used to target and monitor restoration sites. The program has been transferred to several other estuarine locations throughout the U. S. and is considered a model program for public involvement.

Introduction

The importance submerged aquatic vegetation (SAV) species to estuarine and coastal fisheries productivity has long been recognized. In recent decades, dramatic losses of SAV have been documented throughout its range and attributed to a variety of causes, including water quality degradation resulting from eutrophication (Kemp et al., 1983), aquacultural practices, coastal development (Short and Burdick, 1996), and human-induced disturbance and storm events (Short and Wyllie-Echeverria, 1996). The current consensus among scientists and resource managers is that reduced light availability caused by poor water quality resulting from the pollution associated with increased human population and development is the most important cause of SAV loss (Duarte, 1995; Short and Wyllie-Echeverria, 1996; Short and Burdick, 1996).

On average, SAV species require approximately 10.8% of surface light to survive (Duarte, 1991). The amount of light available underwater decreases

[1] Watershed Restoration Program Director, Alliance for the Chesapeake Bay, 6600 York Road, Suite 100, Baltimore, MD 21212
[2] Citizen Monitoring Program Coordinator, Alliance for the Chesapeake Bay, 6600 York Road, Suite 100, Baltimore, MD 21212

exponentially with depth due to the scattering, reflection, refraction, and absorption of incident light caused by the water itself and dissolved and particulate constituents within the water column (Dennison et al., 1993). Recently, research and management efforts have focused on identifying which water quality parameters most affect light extinction and are correlated with seagrass survival (Batuik et al., 1992; Stevenson et al., 1993; Morris and Tomasko, 1993; Fletcher and Fletcher, 1995). In the Chesapeake Bay, the critical water quality parameters have been identified as total suspended solids, dissolved inorganic nitrogen (DIN), dissolved inorganic phosphorous (DIP), and chlorophyll *a* concentrations (Batuik et al., 1992; Stevenson et al., 1993; Dennison et al. 1993). In other areas, such as the Indian River Lagoon of Florida, water color (Kenworthy and Fonseca, 1996) and dissolved organic matter have been shown to be important. In the Long Island Sound, the importance of tidal range and sediment organic matter content have also been documented. Based on the results provided by these studies, many estuaries now have management plans in place to improve the quantity and quality of light reaching the bottom, primarily through a reduction in the amount of nutrient pollution, such as DIN and DIP, reaching the waterbody (Batuik et al. 1992; Morris and Tomasko, 1993; Duarte, 1995).

The reduction of nutrient pollution is expected to improve water quality and increase light availability, allowing SAV species to recover. Some SAV species have dramatic effects on physical and biological site characteristics and water quality. For example, eelgrass (*Zostera marina* L.) plants provide sediment stabilization (Ward et al., 1984; Hine et al., 1987), and baffle wave energy (Gambi et al., 1990), creating a depositional environment (Ward et al., 1984). SAV species also filter and retain nutrients from the water column (Kenworthy et al., 1982; Short and Short, 1984). SAV species support large and diverse faunal assemblages, often with a different species composition than that found in unvegetated areas (Orth et al., 1984; Heck et al., 1995). Consequently, once SAV cover is lost, research has shown that physical and biological site characteristics and water quality change. Rasmussen (1977) and Christiansen et al. (1981) documented the subsidence and loss of fine particle sediments and organic matter with the disappearance of eelgrass in Denmark following the wasting disease of the 1930's. Hine et al. (1987) reported an increase in sediment transport and decrease in sediment deposition associated with the loss of seagrass along the Florida coast. Short term water quality degradation caused by sediment resuspension was reported by Olesen (1996) and Duarte (1995) in areas where seagrass cover was lost.

The presence of SAV species have also been shown to have a significant influence on faunal recruitment processes (Grizzle et al., 1996; Eckman, 1987) and predator-prey relationships (Heck and Crowder, 1990; Orth et al., 1984), suggesting that the loss of SAV can result in significant changes to the biological communities inhabiting a site. These types of physical and biological site characteristics may be important determinants for the potential recolonization of historically vegetated sites, even after sufficient water quality improvements have been made.

Project Description

The Alliance for the Chesapeake Bay's Citizens Monitoring Program (CBCMP) has been a national model for generating quality assured data collected by volunteers. The program has become an important tool for augmenting water quality data for the Chesapeake Bay Program and providing a useful means for public involvement. In 1994, the Alliance for the Chesapeake Bay began a three year pilot program to investigate if citizens could collect data and prepare samples to characterize the habitat requirements for SAV species in select tributaries of the Chesapeake Bay. After three years of data collection, 1995 - 1997, the pilot program has generated quality assured data that has been used to target and monitor restoration sites. During the pilot stage, the scope of the project has expanded in terms of sites and in terms of application of the data generated to target sites with restoration potential and to use as a tool to monitor restoration sites.

The pilot program was initially conducted to determine the feasibility of including the citizen-based monitoring of SAV habitat requirements as part of long-term Chesapeake Bay monitoring efforts to acquire quality assured data that links key water quality parameters with living resources. A major goal of this effort is to advance the understanding of the relationship of watershed management and the distribution of SAV. Citizen Monitoring data is made available to resource management agencies, planners, and local, state and federal government decision makers as well as the monitors themselves.

At each monitoring station, trained citizens measured the following water quality parameters considered to be essential habitat requirements for SAV (Batiuk et. al. 1992): Light attenuation (Kd); Dissolved inorganic nitrogen (DIN)Dissolved inorganic phosphorus (DIP); Total suspended solids; and Chlorophyll *a*. Volunteers undergo a 4 hour training session covering sampling procedures, safety and quality assurance. Training continues at a field sampling station where volunteers are asked to conduct the sampling procedures under careful review. Oversight is provided throughout the monitoring period with occasional visits by Alliance staff to maintain communication with volunteers. Results are reported to volunteers to reinforce quality control, provide an educational experience and to instill a sense of being a vital part of the project team

Light penetration within the water column is measured by using a 20 cm. Secchi disk with the measure of depth later converted to the light attenuation coefficient (Kd). Nutrient concentrations, turbidity and Chlorophyll-a concentrations are measured with the use of a filter apparatus and samples are prepared for laboratory analysis. Data sheets are used to record the time of day, approximate cloud cover, volume of water filtered and Secchi depth and PAR meter readings, if applicable. Upon completion of the day's sampling, a copy of the data sheet is sent to the Alliance for verification of regular sampling. A sample custody sheet which contains the station locations and volumes filtered remains with the samples that are immediately frozen for later shipment to the lab.

At regular intervals, not to exceed 28 days, the samples prepared by the volunteer participants are delivered to the laboratory for analysis. Dissolved inorganic nitrogen,

DIN , is measured by the lab through the concentration of nitrate (NO_3^-), nitrite (NO_2^-), ammonium (NH_4^+). Measurements of orthophosphate (PO_4^{3-}) is used to calculate the concentration of dissolved inorganic phosphorus, DIP. Concentrations of chlorophyll *a* are conducted using the spectrophotometric method (ASTM Method D3731-79). Turbidity is measured through a weighing process of a filter that has been prepared by the volunteer monitors. The concentration of suspended solids is determined in mg/l.

Results

From 1995 through 1997, volunteers collected data on the five habitat requirements every two weeks throughout the growing season for an average of 14 site visits per year. The data generated for each habitat requirements during an entire growing season is calculated into a single median value. Each of the five median values are presented in terms of the habitat requirement breakpoints established by Batiuk et. al. 1992. The results of the three year demonstration phase indicate that citizens were able to follow strict quality assurance/quality control procedures, willing and able to follow the bi-monthly data collection schedule, and the public involvement aspect raised awareness about the importance of SAV and improved water quality.

The results suggested that citizen monitoring of SAV habitat requirements could be a cost-effective method to improve characterization of physical and chemical parameters and their relationship to living resources as well as an efficient means to public involvement in the restoration of the Chesapeake Bay. A review of the performance of volunteer participants who monitored the sites during the three year pilot project indicate that they value the involvement in the program and can be expected to perform the quality assured data collection techniques. However, it may be unrealistic to expect this data to be used to fully characterize the habitat requirements along an entire tributary from data collected at three or four near shore stations. Pulses in turbidity, nutrients, algae growth, tide conditions, and periods of high wave energy all exist in each tributary and a grab sample every two weeks leading to an annual median value will not likely accurately reflect these temporal conditions.

Conclusion

The monitoring procedure has been shown to be effective in characterizing the habitat requirements at a specific location. From the results of this data, the Alliance for the Chesapeake Bay suggests that this monitoring protocol be best used to target suitable restoration locations for direct transplanting of SAV as a means of achieving SAV restoration goals as set by the Chesapeake Bay Program Executive Council.

Literature Cited

Batiuk R.A. et. al. Submerged Aquatic Vegetation Habitat Requirements and Restoration Targets: A Technical Synthesis. U.S. E.P.A. for the Chesapeake Bay Program. Annapolis, MD. December 1992.

Christiansen, C., H. Christofferson, J. Dalsgaard, P. Nornberg, 1981. Coastal and nearshore changes are correlated with die-back in eelgrass (Zostera maina). Sed. Geol. 28: 168-178.

Duarte, C.M. 1995. Submerged Aquatic Vegetation in Relation to Different Nutrient Regimes. Ophelia 41: 87-112.

Eckman, J.E. 1987. The role of hydrodynamics in recruitment, growth, and survival of *Argopecten irradians* (L) and *Anomia simplex* (D'Orbigny), within eelgrass meadows. J.Exp. Mar. Biol. Ecol. 106:165-191.

Fonseca, M.S. 1992. Restoring seagrass systems in the United States. Pp.79-110. In G.W. Thayer (ed.), Restoring the Nation's Marine Environment, Maryland Sea Grant, College Park, Maryland.

Fletcher, S.W. and W.W. Fletcher. 1995. Factors affecting changes in seagrass distribution and diversity patterns in the Indian River Lagoon complex between 1940 and 1992. Bull. Mar. Sci. 57 (1): 49-58.

Gambi, C.M., A.R.M. Nowell. And P.A.Jumars. 1990. Flume observationson flow dynamics in *Zostera marina* L. (eelgrass) beds. Mar. Ecol. Prog. Ser. 61:159-169.

Grizzle, R.E., F.T. Short, C.R.Newell, H. Hoven, and L. Kindbloom. 1996. Hydrodynamically induced synchronous waving of seagrass "monarni" and its possible effects on larval mussel settlement. J. Exp. Mar. Biol. Ecol. 206:165-177.

Heck, K. L. and J.F. Valentine. 1995. Sea urchin herbivory: evidence for long-lasting effects in subtropical seagrass meadows. J.Exp. Mar. Biol. Ecol. 189:205-217.

Heck, K.L. and L.B. Crowder. 1990. Habitat structure and predator-prey interactions. *In:* Bell, S., E. McCoy, and H. Muchinsky (eds.), Habitat complexity: the physical arrangement of objects in space, pp. 281-299. Chapman and Hall, New York.

Kenworthy, W. J., J.C. Zieman and G. W. Thayer. 1982. Evidence for the influence of seagrasses on the benthic nitrogen cycle in a Coastal Plain estuary near Beaufort, North Carolina (USA). Oecologia, 54:152-158

Morris, L. J. and D.A Tomasko (eds.). 1993. *Proceedings and conclusions of workshops on: Submerged aquatic vegetation and photosynthetically active radiation.* Special Publication SJ93-SP13. Palatka, Fla.: St. Johns River Water Management District.

Olesen, B. 1996. Regulation of light attenuation and eelgrass *Zostera marina* depth distribution in a Danish embayment. Mar. Ecol. Prog. Ser. 134:187-194.

Orth, R. J. and J. van Montfrans. 1984. Epiphyte-seagrass relationships with an emphasis on the role of micrograzing: a review. Aquat. Bot. 18:43-69.

Rasmussen, E. 1997. The wasting disease of eelgrass (Zostera marina) and its effects on environmental factors and fauna. In: McRoy, C.P. and C Helfferich (eds.) Seagrass ecosystems: a scientific perspective. Marcel Dekker, New York, p. 1-52.

Short, F. T. and C. A. Short. 1984. The seagrass filter: purification of estuarine and coastal waters. In V. S. Kennedy (Editor) The Estuary as a Filter. Academic Press, pp 395-413.

Short, F.T., D.M. Burdick, S. Granger, and S.W. Nixon. 1996. Long-term Decline in Eelgrass, *Zostera marina* L., Linked to Increased Housing Development. Seagrass Biology: Prodeedings of an International Workshop, 291-298.

Short, F.T., and S. Wyllie-Echeverria. 1996. Natural and Human-induced Disturbance of Seagrasses. Environmental Conservation, 23(1): 17-27.

Stevensen, J.C., L.W. Staver, and K.W. Staver. 1993. Water Quality Associated with Survival of Submersed Aquatic Vegetation Along an Estuarine Gradient. Estuaries, Vol.16, No.2: 346-361.

Ward, L.G., W.M. Kemp and W. R. Boynton. 1984. The influence of waves and seagrass communities on suspended particulates in an estuarine embayment. Mar. Geo., 59:85-103

GREAT LAKES

Moderator, Don Vonnahme

"Interstate Mediation as an Alternative to Litigation"
Daniel Injerd
Illinois Department of Natural Resources

"Chicago Sanitary and Ship Coral Dispersal Barrier"
Phil Moy
Corps of Engineers, Chicago

"A Long Term Strategy for Sediment Placement"*
John Loftus
Toledo-Lucas County Port Authority

"The Pulp and Paper Industry and the Great Lakes Water Quality Initiative"
Dale Phenecie
Environmental Affairs Consulting

"The Economic and Environmental Aspects of Nonpoint Pollution Control"
James Ridgway
Environmental Consulting and Technology, Inc.

*Text not available at time of printing.

Interstate Mediation As An Alternative to Litigation: Resolving Illinois' Lake Michigan Diversion Dispute

Daniel Injerd[1]

Abstract

In 1994, Illinois was notified that it had violated the United States Supreme Court Decree that limits Illinois' diversion of water from Lake Michigan. As an alternative to litigation, the Great Lakes states and the U.S. Department of Justice utilized non-binding mediation to resolve this dispute. A mediator with experience in interstate water disputes was retained, and the first of four mediation sessions was held in October 1995. On October 9, 1996 the Great Lakes states announced the approval of a Memorandum of Understanding which sets forth a process and establishes key benchmarks, which, if completed over a 3 year period, will enable the parties to petition the Supreme Court to amend its previous decree and bring this century old diversion dispute to a final and comprehensive resolution.

Background

Illinois' diversion of water from Lake Michigan dates back to 1848, when the Illinois and Michigan (I&M) Canal was opened to traffic. The I&M Canal established a waterway connection between the Great Lakes and the Illinois and Mississippi River system. With the completion of the Chicago Sanitary and Ship Canal in 1900, diversion capacity was greatly increased from 2.8 cubic meters per second (cms) (100 cubic feet per second, cfs) to 280 cms (10,000 cfs). Annual average diversion peaked in the late 1920's at close to 280 cms (10,000 cfs), then began decreasing. From 1938 to the present, Illinois' annual average diversion has remained relatively constant at 90 cms (3,200 cfs).

---

[1]Chief, Lake Michigan Management Section, Office of Water Resources, Illinois Department of Natural Resources, 310 S. Michigan Avenue, Room 1606, Chicago, Illinois 60604

Illinois' diversion of water from Lake Michigan has generated sufficient controversy among the Great Lakes states and lower Mississippi River states such that the dispute has been before the U.S. Supreme Court on numerous occasions over the past century. In 1967, the Court approved a new decree that limited Illinois' diversion to 90 cms (3200 cfs) including domestic pumpage. In 1980, the Court amended the 1967 Decree to extend the averaging period from 5 to 40 years to allow Illinois to use its diversion more efficiently.

Illinois' diversion consists of 3 primary categories: domestic water supply, direct diversion into the Chicago Sanitary and Ship Canal and stormwater runoff from the diverted watershed. Domestic water supply is the largest, with 6.3 million people currently relying on the lake for their water supply.

Lake Michigan water is diverted directly into the Chicago Sanitary and Ship Canal to provide for safe navigation and to improve water quality in the canal.

The last category of Illinois' diversion is stormwater runoff from the 1750 km^2 (673 mi^2) watershed that was diverted by the reversal of the Chicago and Calumet Rivers.

Diversion Accounting

Determining Illinois' diversion is an extremely time consuming and complicated project. It is an archaic system, owing to its derivation from a succession of Supreme Court Decrees. The 1980 amendments to the 1967 Decree retained the historic diversion accounting system, but in addition to lengthening the averaging period from 5 to 40 years, also directed that the accounting system utilize the "best scientific knowledge and engineering practice." Illinois took that directive seriously, and in 1983 implemented major improvements in diversion accounting. The most significant improvement was the installation of an acoustic velocity flowmeter to measure total flow in the Chicago Sanitary and Ship Canal. Results from this new gage indicated that the old measurement system was underreporting flows. Certified diversion accounting reports show a distinct increase in Illinois' diversion, starting in 1983, though there was no increase in Illinois' use of Lake Michigan water. In 1984, the Corps of Engineers (Corps) assumed the operation and maintenance responsibility for the Chicago River lock, and in 1986, responsibility for performing the measurements and computations to determine Illinois' diversion was transferred from the state to the Corps.

In 1990, Illinois notified the Corps that the Chicago River lock gates were not closing completely, allowing a significant amount of Lake Michigan until May 1993. The Corps' diversion accounting reports included this leakage flow as diversion by the State of Illinois.

Mediation

In January 1994, Illinois, along with the other seven Great Lakes states, was notified by the Corps that it had violated a diversion limit in the 1980 Decree. The principal causes for this violation were: (1) improvements in the diversion accounting measurement technology implemented by the State of Illinois revealed that the historic diversion accounting system, upon which the Decree limit of 90 cms (3200 cfs) was based, in fact underreported actual diversion by approximately 8.4 cms (300 cfs), and (2) excessive leakage through the Chicago River lock, a Corps' facility, accounted for an additional 7-8.4 cms (250-300 cfs) of diversion. The Corps concurred with this assessment, but was not willing to either adjust the diversion accounting process to reflect the new measurement system or to exclude from Illinois' diversion the excess leakage through diversion structures under the jurisdiction and control of the federal government.

Both Wisconsin and Michigan responded to the Corps report. In March 1995, the Attorney General of Michigan brought this matter to the attention of the Chief Deputy Clerk of the U.S. Supreme Court. In response to the concerns expressed by these states, the U.S. Department of Justice scheduled a briefing for interested states on June 15, 1995. From the very beginning, the Justice Department expressed a strong desire for the parties to the original Supreme Court litigation to reach an accommodation without costly and time consuming litigation. Their experience with original jurisdiction cases involving water disputes between states is that a 10 year timeframe and a $10 million cost is the norm, so they highly recommended the use of an alternative dispute resolution mechanism to resolve this dispute. Fortunately, after a round of initial posturing, the states agreed to use a non-binding mediation process to attempt to resolve the dispute over Illinois' alleged overdiversion. In October 1995, the Great Lakes states and the Department of Justice selected a mediator to assist the states in resolving their differences and to try and reach a settlement outside of the Court.

The first formal gathering of the states, the federal government and the mediator occurred on October 23, 1995. At this first session, the parties agreed to several ground rules that set the stage for productive discussion. These included a confidentiality agreement so that the parties would be free to share their positions on issues with the mediator, that the mediator would handle all press contacts, that the cost of the mediator would be equally borne by all parties, including the federal government, and that each party would promptly submit to the mediator a confidential statement setting forth their claims, position and contentions in this dispute.

Each state was free to select the members of their mediation team. The contact list for this mediation contained 45 names; at times, Illinois' mediation team had up to 15 members. Members of Illinois' mediation team included the Attorney General's office, legal and senior policy staff from the Illinois Department of Natural Resources, senior policy staff from the Illinois Environmental Protection Agency, legal and senior staff from the Metropolitan Water Reclamation District of Greater Chicago and legal and senior staff from the City of Chicago. All of the other Great Lakes states had at least one legal representative and one technical representative participating in the mediation process.

Early on in the mediation, a list of technical and legal issues were identified. They included:

- Is Illinois in compliance with the United States Supreme Court Decree, as amended in 1980?
- If not, is the reason due to leakage and to a change in measurement procedures? Are these valid reasons?
- Has the diversion increased over time?
- Should leakage through diversion structures under the jurisdiction of the U.S. Army Corps of Engineers be charged to Illinois?
- Should the 1980 Decree be modified?
- If Illinois is out of compliance, is it feasible/possible for Illinois to come back into compliance? What conservation measures (i.e., Chicago metering) can Illinois implement to reduce diversion? How much will it cost, and how long will it take?
- What are the consequences of reducing discretionary (water quality) flows in the canal system?
- How can the Decree be enforced?

These issues were discussed and debated in great length over the next 10 months at 3 mediation sessions, each lasting 2 days. These sessions were long, grueling affairs and involved both times when the whole group met and when two or more groups would go to separate rooms and the mediator would shuttle back and forth, delivering positions, responses, and, perhaps most importantly, identifying areas of agreement or acceptance.

Summary of Memorandum of Understanding (MOU)

On October 8, 1996, the governors of the 8 Great Lakes states and the Department of Justice announced the signing and acceptance of the Great Lakes Mediation Memorandum of Understanding (MOU). This MOU set forth both a process for resolving the dispute through amending the Decree and also includes very specific tasks that the State of Illinois must complete over a 3 year period to reduce diversion from Lake Michigan and demonstrate the validity of a new lakefront diversion accounting system. The key elements of the MOU include:

- Illinois will comply with the diversion limit of 90 cms (3,200 cfs) and will restore the excess diversion it has withdrawn since 1980.
- The diversion measurement system will be moved to the lakefront with the goal of improving the accuracy and timeliness of the reporting system. A stormwater runoff value of 22 cms (800 cfs) will be used, along with a consumptive use adjustment of 5 cms (168 cfs), to develop a lakefront diversion limit of 72 cms (2,568 cfs).
- The State of Illinois will provide the non-federal cost to the United States Geological Survey to install state-of-the-art acoustic velocity flowmeters to measure direct diversion in the Chicago and Calumet Rivers.
- Over a 3 year transition period, the Corps of Engineers will maintain a dual accounting system to compare the new lakefront accounting system to the current accounting system as directed in the 1980 amended Decree.
- To reduce diversion, Illinois will reduce discretionary diversion in the Chicago Sanitary and Ship Canal subject to water quality maintenance, by December 1998 will initiate construction of a new wall across the mouth of the Chicago River Turning Basin to reduce leakage, by October 1, 1998 will install one or more pumps at the mouth of the Chicago River to return any remaining leakage flows back to Lake Michigan, and will review all its water allocations and pursue enforcement of efficient water use and conservation measures.
- During the transition period, no party will initiate legal action. If Illinois lives up to its commitments and the accuracy of the lakefront accounting system is demonstrated, then the parties will pursue an amended Decree that incorporates the terms agreed to in the MOU.

Lessons Learned

Mediation works! This water diversion dispute, as evidenced by its long history at the Supreme Court, has remained an extremely controversial issue, and carries a unique set of technical, legal and political challenges. The fact that a group of lawyers, engineers and technical and policy staff from 8 Great Lakes states could sit together around a table and reach consensus on this issue in just one year is truly nothing short of remarkable. As a first time observer of a formal mediation process, I think there were a number of items that contributed to a successful mediation process. These include:

- The alternative to mediation, litigation at the Supreme Court, is a strong incentive for mediation.
- The U.S. Department of Justice was a strong proponent of alternative dispute resolution, and their encouragement and assistance in this effort was invaluable.
- Involving all parties to the dispute in the selection of a mediator gave the mediator instant credibility. Having an experienced and skilled mediator was extremely important, and allowed the mediation process to move forward.
- Participating in the mediation process was a long, tedious process, and all participants experienced times of frustration and impatience. However, the investment of time and energy by the participants created a desire to achieve a successful mediation.
- Initial positions that were seen as "deal breakers" were not in fact "deal breakers". As the mediation process moved forward, parties discovered their positions were more flexible than they initially believed.
- Legal and technical staff can interact much more productively in a mediation process, where both sit at the table, rather than in litigation.
- Make sure that the needs of the mediation participants are not neglected. Stay at a nice place, and make sure participants are well fed and cared for.

Summary

There is no question that the use of mediation enabled this dispute to be kept out of the Supreme Court. That mediation was successful in resolving a very complicated water diversion dispute, with a long legal history, political overtones and detailed technical issues, indicates that mediation can be used in almost any circumstance. For this mediation process to remain a success, though, requires that over the next several years Illinois invest between $15-20 million in projects that will reduce diversion and continue its commitment to the wise use and management of our limited diversion, and the Corps must maintain both accounting systems and demonstrate the reliability of the lakefront system. Illinois is committed to fulfilling its obligations and is confident that at the completion of the transition period the parties will be able to petition the Court to amend its previous decree and bring this century old diversion dispute to a final and comprehensive resolution.

Chicago Sanitary and Ship Canal Dispersal Barrier

Philip B. Moy[1]

Abstract

Inter-basin dispersal of aquatic nuisance species poses a threat to the integrity of our native and managed ecosystems. The Chicago Sanitary and Ship Canal and Calumet-Saganashkee Slough (Cal-Sag) Channel, which provide an important trade corridor, now also form the sole aquatic link between the Lake Michigan and Mississippi River basins. Following the zebra mussel, other invasive species may use this corridor for range expansion. Species in the Great lakes such as the round goby, white perch, three-spine stickleback and ruffe may spread downstream into the Des Plaines, Illinois and Mississippi Rivers and their tributaries. This is the path the zebra mussel followed into the Midwest. Likewise, introduced species in the Mississippi River basin can use this aquatic pathway to gain entry to the Great Lakes. Bighead carp, black carp, grass carp, striped bass and its hybrid with the white bass, are headed towards the Great Lakes. Fish are not the only problem; zooplankton forms the first forage for many of our prized sportfish. Large, spiny zooplankton species from Europe and Africa compete with our native zooplankton and form a non-consumable portion of the forage base for small fish.

Concern for the dispersal of these organisms between these two great natural drainage basins focused attention on the linkage between them formed by the Sanitary and Ship Canal. There is great interest in creating a barrier to prevent the dispersal of these aquatic organisms between the Great Lakes and Mississippi River basins.

----

[1]Fisheries Biologist, U.S. Army Corps of Engineers Chicago District, 111 North canal Street, Chicago, IL 60606-7206

Recognizing the complexity of the system and the diverse uses and activities on the Sanitary and Ship Canal, the Corps of Engineers assembled an advisory panel comprised of waterway users, riparian owners, regulators and biologists. Federal member entities include the US Army Corps of Engineers Chicago District and Waterways Experiment Station, the US fish and Wildlife Service, the US Geological Survey Biological Resources Division and the US Environmental Protection Agency. State members include the Illinois Department of Natural Resources, the IDNR Office of Water Resources, the Illinois Natural History Survey, the Illinois Environmental Protection Agency and the Illinois Pollution Control Board. Regional entities include the Metropolitan Water Reclamation District of Greater Chicago, DuPage County Forest Preserve Illinois-Indiana Sea Grant, the City of Chicago Mayor's office, and the Northeastern Illinois Planning Commission. The Illinois River Carriers Association, Illinois International Port and Commonwealth Edison are also represented. Two environmental groups are involved: the Friends of Chicago River and the Great Lakes Sport Fishing Council. The panel also has participants from Loyola University, University of Michigan and the University of Windsor.

Through a series of meetings the panel quickly identified uses of the canal which should not be affected by creation of a dispersal barrier: commercial shipping and the Lake Michigan diversion volume. The uninterrupted passage of barge traffic is considered one of the paramount uses of the waterway. Likewise, passage of treated sanitary discharge and the Lake Michigan diversion water could not be changed. Other obstacles to development of a dispersal barrier are the variation in water flow volumes (2000 to 20,000 cubic feet per second), recreational boating and public perception.

Each participating entity brings different expertise, authorities or concerns to the table. The participation of the permitting agencies has been particularly beneficial; it has allowed identification of potential permitting concerns at an early stage. The Illinois River Carriers and the Metropolitan Water Reclamation District have been able to point out potential conflicts with certain barrier approaches. The researchers familiar with fish behavior and biology have been able to propose options for the barriers and provide immediate input as the barrier concepts were modified to accommodate potential on-site conflicts.

It was critical to achieve a consensus among these participants on a potential and acceptable approach for a dispersal barrier and the ways and means to implement it. Participants agreed that a barrier was needed, but they were divided on how best to achieve the goal. Potential approaches considered for the barrier were:

Chemical - rotenone, antimycin, chlorine
Physiological - heat, low dissolved oxygen, nitrogen stripping
Visual - light, bubbles
Acoustic - high, low frequency sound

The visual and acoustic barriers were not considered to be effective enough to be used alone, but may be useful in conjunction with other approaches. The reversibility or detoxification of a given method was an important factor in the approach to the barrier. The ideal barrier would stop the passage of organisms, allow shipping and sanitary discharge to take place, be low cost and the effect would be limited to a discrete reach of the canal. Operating cost, existing water quality permit requirements and availability of existing technology were also factors used to rank the approaches. Of the possible approaches considered, an electric field and rotenone (a fish toxicant) were considered most promising; chlorine was a close third. The advisory panel recommended that the use of toxicants be limited to preemptory measures only, rather than as an ongoing treatment.

The use of an electric barrier was indicated as the most likely approach for ease of development in the Sanitary and Ship Canal. The presence of the goby in the Cal-Sag Channel, only about 17 miles upstream of the confluence with the Sanitary and Ship Canal, provided the impetus to develop a benthic barrier for this bottom dwelling fish. This prompted a phased approach to the development of the dispersal barrier. The first phase would consist of a benthic barrier targeting gobies. If that proves successful, phase two would involve moving to a full water column electric barrier for active swimming fish. The final phase will eventually consist of a barrier for planktonic forms. It is likely that the most effective and reliable barrier will be comprised of a combination of approaches. This could involve acoustic barriers and or visual barriers possibly combined with a zone of low dissolved oxygen which fish will voluntarily avoid.

Several agencies are cooperating to fund development of the barrier. The US Army Corps of

Engineers has been provided $500,000 and the US Environmental Protection Agency has been provided $250,000 for investigation of the barrier. The US Geological Survey - Biological Resources Division has provided $30,000 to examine goby response to the electric barrier in the laboratory. The Great Lakes Protection Fund has contributed $71,000 for monitoring, and the Illinois International Port has contributed $75,000 towards construction of the barrier.

Developing and establishing this barrier is vital to maintaining the health of Midwest ecosystems, particularly in the Mississippi River. As the project progresses, we anticipate additional Mississippi basin states will be become involved as participants in the advisory panel and as contributors as well.

The Pulp and Paper Industry and the Great Lakes Water Quality Initiative: Achieving Environmental Compliance in a Competitive Industry

Dale K. Phenicie[a]

Abstract

Water resource management is a critical matter for the pulp and paper industry. Large quantities of process water are needed. Large quantities of treated effluent must be properly released back into surface waters. Regulatory controls placed on the discharge permit process can have substantial technological and economic impacts on pulp and paper mill operations. Representatives of the pulp and paper industry have followed the development of U.S. EPA's Great Lakes Water Quality Initiative (the GLI)$_{(3)}$ closely. The GLI will change, substantially, the permitting process and regulatory outcome for, not only paper mills, but all dischargers in the Great Lakes Region. To the extent that these new policy directives are applied elsewhere, the process will change in other locals as well. This paper discusses the significance of the regulatory process changes brought about by promulgation of the GLI. A case study which identified potential GLI rule permit application cost increases, incurred in support of a hypothetical permit modification, for a POTW serving a recycled paper mill is reviewed.

Introduction

Water is an essential resource for the manufacture of pulp and paper products. It serves as the medium or base, within which chemical reactions needed to separate lignins from wood fibers take place. It supports pulp bleaching or whitening chemistry, and serves as the essential reagent required to promote hydrogen bonding between fibers, thus imparting strength to finished paper products.

Modern process water recycling technologies have drastically reduced the amount of fresh water required for the production of each ton of product by over 70 percent since 1959$_{(1)}$. However, most pulp and papermaking processes still require the use of fresh water, and result in the need to treat and discharge wastewater to a receiving

---

[a] Environmental Affairs Consulting, 402 Lighthouse Lane, Peachtree City, GA, 30269

stream. The release of this treated effluent is regulated through the United States National Pollutant Discharge and Elimination System (NPDES) permitting process. U.S. EPA, or state agencies to which permitting authority has been delegated, utilize regulatory standards and set limits regarding allowable pollutant concentrations for the final effluent discharged at each mill.

In March of 1995, through promulgation of the GLI, U.S. EPA established a new wastewater discharge permitting protocol for NPDES regulated dischargers located within the Great Lakes Basin. This new set of permitting standards and procedures has forever changed the processes under which discharge limits are to be set for Basin industries and municipalities. Because of the need for continued use of large quantities of water, and the nature of the new permitting process, the pulp and paper industry is particularly impacted by the GLI.

A Comparison of Permitting Processes

For pulp and paper dischargers, as with others, the typical NPDES permitting process has been centered around conventional pollutants such as total suspended solids (TSS), biochemical oxygen demand (BOD), chemical oxygen demand (COD), pH, color, turbidity, and temperature. A checklist style review of the potential for the presence of “priority” or “toxic” pollutants was included in the process, along with a requirement that effluents meet toxicity standards, relative to the survival of selected test species, through whole effluent toxicity (WET) testing protocols. In addition, attention was given to receiving stream conditions and the need for special discharge limitations to address water quality limited circumstances. This process required the use of historical data to characterize the effluent stream in terms of conventional pollutants, some knowledge of or perhaps limited testing for, the presence of priority pollutants, evaluation of effluents using WET testing protocols, and knowledge of or examination of receiving stream conditions in the vicinity of the discharge point.

The new GLI regulations change this process substantially. While conventional pollutants will still be addressed in much the same way, a new focus has been added in the “toxics” area. For the first time a very extensive “toxics” protocol will have to be followed. It includes the need to fully evaluate effluents for the presence of approximately 138 chemical substances at extremely low levels. In some cases, biomonitoring techniques or other measures will have to be used to estimate concentrations of chemical substances at levels below analytical detection capabilities. This characterization work is needed in order to determine the potential of the effluent to cause receiving water quality standards, or criteria, for each of the substances to be exceeded. For substances for which receiving stream water quality criteria have not been developed, such criteria will have to be developed by the permitting authority, or the applicant, if the process is to move forward and a new permit granted.

Adding to the complexity of these new requirements if the fact that four separate sets of criteria must be determined for each of the 138 chemicals. Criteria must be determined for human health, wildlife and aquatic species. Two separate aquatic species criteria are required, one for acute and one for chronic effects. Permit limits, for the 138 chemicals, are set on the basis of which of these four criteria turns out to be lowest in each case.

Obtaining new or revised permits can be critical if a significant process modification or plant expansion project is desired. Although the new rule provides an abbreviated water quality criteria (or "tier II value"[b]) setting procedure for use in determining the likelihood of exceedances, this process allows the use of uncertain data and requires additional safety factors which can drive resulting permit limits to unrealistically low levels.

In addition to requiring the extensive "toxics" centered effluent characterization and "potential to exceed" process, the GLI also requires the use of newly formalized Total Maximum Daily Load (TMDL) procedures on receiving stream watersheds, both below and above the permitted discharge outfall point. These detailed procedures will have had to have been previously performed, for the particular chemical substances in question, or will have to be performed before a new permit or permit renewal can be granted. While the TMDL responsibility is that of the permitting authority, the applicant may have to agree to perform or finance some or all of the work in order to obtain a needed permit.

The GLI also limits or eliminates the use of certain permitting tools which have been included in NPDES processes since the first permits were issued nearly 25 years ago. For example, mixing zones for certain chemical pollutants[c] are being phased out. This will require that effluent limits for these substances be set at receiving water quality criteria levels. Once phased out, no dilution effect can be recognized or taken advantage of. As a result, effluent treatment processes or internal manufacturing processes will have to be capable of limiting discharges to these extremely low, and often non-measurable, ambient receiving stream criteria levels. Additionally, credits for pollutants which enter a manufacturing plant process stream with intake process water are subject to tight interpretation guidelines, making them less likely to offset measured or estimated discharge quantities subject to permit limits.

Finally, applications for permits under the GLI regulation must be accompanied by extensive pollutant minimization plans in cases where discharge limits are set at levels below analytical detection capability.

---

[b] "Tier II" wildlife values are only used if endangered species are present.

[c] Bioconcentrating Chemicals of Concern (BCCs)

Aspects of the GLI Rule Responsible for Pulp and Paper Industry Impacts

The pulp and paper industry's need to rely on water based processes, and the fact that a second natural resource, wood, may carry naturally occurring chemical substances into these processes makes the Industry particularly sensitive to GLI regulatory changes. Common substances including mercury, copper, zinc, aluminum, and iron are among the GLI regulated substances which can be brought into mill process through the wood supply. A variety of organic compounds also included in the GLI substances list may also accompany the wood or may be produced as a result by pulping/bleaching processes. In addition, many of these chemicals or compounds also enter mill processes through process water intake and as a result, must be addressed in mill effluent. As explained above, the nature of the GLI standard is such that individual mill permit limits for these materials become set at very low levels.

The GLI rule does include a number of "off ramps" which may provided for limited exemptions from the tight limits. However, most of these variances are temporary and/or require extensive demonstrations which make qualification for them difficult or unlikely. Ultimately, additional advanced treatment systems will probably be necessary for compliance, and costs can be very high. An engineering study during the proposal stage of the GLI regulatory development process estimated treatment system costs for Basin pulp and paper mills at more than one billion dollars (2).

But, regardless of how the implementation process goes for a given mill, other industry or municipality, and/or regardless of which of the several "off ramps" a facility may qualify for, a real and very significant cost will be in the permitting process itself. As explained above. Several new steps have been added to the wastewater discharge permitting process by this rule. It is this change in process which will substantially extend the time necessary to apply for, and obtain, the permit and it will markedly increase permit application costs.

Gary Indiana Case Study

Permit application/processing costs were studied for the Gary Sanitary District in Gary Indiana, a POTW serving a recycled paper mill owned by Georgia-Pacific Corporation (4) . This treatment facility discharges to the Grand Calumet River for which a Waste Load Allocation study had been previously performed. The POTW had recently renewed it's State issued NPDES permit. The premise behind this study was that a subsequent permit modification was needed because of a proposed expansion of the paper mill, one of the POTW's significant users.

Under the existing Indiana permitting program, the following information was required of applicants for new or renewed NPDES wastewater discharge permits.

- Basic discharge description for the influent, effluent and outfalls

- Industrial waste contributions including pollutant scans
- Scheduled improvements
- Maps
- Schematics
- Biomonitoring (WET testing) results.

With GLI regulations in place, additional information must be acquired and filed with the renewal application. These include:

- Receiving stream flow characteristics
  - Harmonic mean flow (for human health criteria comparisons)
  - 30Q5 flow (for wildlife criteria comparisons)
  - 7Q10 flow (for aquatic life criteria comparisons)
- Water column and fish tissue chemical analysis
- Characteristics and chemical analysis for District's 15 outfalls and 7 industrial users
- Characteristics and chemical analysis for all other Grand Calumet point source dischargers
- Characteristics and chemical analysis for all Grand Calumet non point sources discharges.

"Standard" NPDES permit application protocols do not require analysis for 24 of the chemical substances included in the GLI rule. Therefore, additional testing was assumed to be required for each of these sources. Additionally, receiving water criteria have been developed for about 13 percent of the GLI listed materials. Hence, if a permit is to be obtained on an expedited basis to support the expansion project, the applicant may have to see that the additional criteria are developed. Since human health and wildlife criteria are established using bioaccumulation factors for listed chemicals, these too would have to be developed.

Once criteria values have been established, the potential to exceed these ambient water quality limits, by each of or by a combination of the discharges, to the Grand Calumet must be determined. For those chemicals for which the criteria are expected to be exceeded, waste load allocation studies must be conducted. Finally, using waste load allocation study results, the developed criteria, and the expected Gary Sanitary District discharge quantities for the GLI listed substances, final permit limits can be calculated. The State, the District and the users will share in the responsibility to support this extensive permitting activity, but ultimately a majority of the cost may have to be borne by the user requesting the increase in flow if an important expansion project is to be supported.

In this case study, the costs and time for collecting the additional information, for conducting necessary analysis, and for preparing permit application documents were determined. They are summarized in the table 1. Under the assumptions used for this case study, costs to prepare permit application materials may cost as much as $3 million. The time required to do the studies, prepare the materials and allow for regulatory review and permitting may take nearly four years.

The Great Lakes States are in the final stages of implementing their new wastewater permitting programs under the GLI. The extent to which applicants can qualify for variances and "off ramps" has yet to be determined. Certainly these provisions and other measures of flexibility, which the states and EPA apply to the new procedures, can reduce the impact of the rule. However, application for, agency review of, and scientific studies necessary to support the "off ramps" will be substantial. Ultimately, Just as the Gary Sanitary District case study has reflected, the permitting process will be lengthy and expensive. To the extent that similar permitting standards do not apply in other parts of the U.S., Great Lakes Basin pulp and paper mills, and other facilities which depend on comprehensive wastewater discharge permits, will find the GLI requirements to be a significant competitive disincentive.

Table 1
Costs and Time Requirements Associated with GLI Style Permit Application Support Documents

Task/Item	**Cost**	**Time Required (months)**
Sampling and analysis of GSD outfalls and user inputs	$146,000	5
Collect other user background date and nonpoint source information	13,500	4
Establish dilution factors (for non BCCs) and Grand Calumet Basis flow characteristics	6,375	4
Develop receiving water quality criteria and Tier II values	2,585,800	12
Estimate expected effluent quality and potential to exceed	6,000	2
Determine Waste Load Allocations	41,250	6
Prepare pollutant minimization program	207,900	3
Prepare final application document package	9,750	3
Prepare applicable variance requests	30,000	3
Agency review, permit drafting and comment period		24
Total	**$3,046,575**	**47[d]**

References

1. American Forest and Paper Association, Washington, D.C., Pollution Prevention Report.
2. American Forest and Paper Association, Washington, D. C. Comments to EPA regarding the proposed Great Lakes Water Quality Initiative.
3. U. S. EPA GLI regulation, 40 CFR Parts 9, 122, 123, 131, and 132.
4. Vorissis, Mary L., Johnson, Pamela D., and Phenicie, Dale K., Great Lakes Water Quality Guidance Impact on Gray sanitary district's NPDES Permit, Conference Proceedings, Water Environment Federation, October, 1994.

[d] Time periods and activities for some tasks are concurrent. Column is not intended to produce an arithmetic total.

The Economic and Environmental Aspects of Non-point Pollution Control: The Rouge River National Wet Weather Demonstration Project

James W. Ridgway, P.E.[1]

Abstract

Over the past 25 years, the Clean Water Act has successfully targeted point sources as the primary pollutant control issue. Increasingly, it has become apparent that non-point sources in many areas are sufficiently large to preclude the desired uses of waterways. Unfortunately, there has yet to be implemented an effective institutional means of controlling these sources. Thus, while a great deal of information has been assembled pertaining to the technical aspects of non-point source control, the actual implementation remains a challenge. The practical aspects of regulating these sources are daunting. This paper summarizes the progress made to date by the Rouge River National Wet Weather Demonstration Project (Rouge Project.) The Rouge Project is supplementing the existing regulatory program with an innovative regulatory program — a general permit for a subwatershed.

Background

In the 25 years of the Clean Water Act of 1972 (PL 92-500), much has been done to remove pollutants from our nations waterways. Unfortunately, urban waterways have yet to achieve the primary goal of the act: fishable and swimmable waters. The Environmental Protection Agency (EPA) and the state regulatory agencies have remained vigilant in enforcing the act when industries and/or municipalities were identified as sources of pollution. This enforcement mentality becomes far less effective when non-point sources became the primary source. Stated simply, there would never be enough environmental enforcement agents to address the millions and millions of small non-point sources.

---

[1]Vice President, Environmental Consulting & Technology, Inc., 220 Bagley Avenue, Suite 600, Detroit, Michigan 48226

Thus, an alternative to "command and control" has been identified as the most practical approach to achieving the goals of the Clean Water Act. The Rouge Project has explored ways to integrate the various federal, state and local statutes and regulations to improve water quality in the Rouge River. It also identified barriers inherent in those existing regulatory frameworks and recommended strategies to overcome them. The goal is to comprehensively protect a watershed that covers multiple political jurisdictions and is threatened by a wide range of pollutant sources. To achieve this goal, the Rouge Project has undertaken efforts to:

- Assist in improving the capability of regional, state and local agencies to address broad programs that affect traditional environmental problems.
- Identify programs that affect resource protection within the watershed.
- Demonstrate existing and proposed interrelationships among institutional stakeholders.
- Identify difficulties and opportunities in integrating diverse programs.
- Increase awareness of new regulations, enforcement authorities, technical guidance and other information affecting environmental management.

Command and Control

The introduction of the Clean Water Act and the National Pollutant Discharge Elimination System (NPDES) program did much to control point sources of pollution throughout the watershed. Unfortunately, many of the gains realized through point source control were offset by the growing non-point loads resulting from uncontrolled urban sprawl. As the downstream communities wrestled with financing combined sewer overflow (CSO) control, many began to ask what level of control is appropriate for a river which will see little or no change in use. This question of equity was further shaded by the fact that the down stream communities are often older, generally less affluent, and made up of a much higher minority constituency. Is it just to ask these communities to finance major CSO control projects with little change in their ability to use the river while the more affluent suburban communities are allowed to continue to transfer significant levels of pollution to their downstream neighbors?

Command and control was extremely successful and we should all be proud of the accomplishments our country has achieved over the past twenty-five years. Unfortunately, it became impractical to control the multitude of non-point sources which contributed to degradation of our lakes and streams. Basically, there could never be enough "environmental cops" to enforce the environmental regulations at each township, subdivision or industrial site. Other pollutant sources which had no apparent owner remained on lists for years but little or no action resulted. Abandoned dumps and contaminated sediments remain a vexing problem in most urban watersheds. Thus, it seems that the strengths of the Clean Water Act should be retained while we look to new ways of solving the problems we cannot seem to overcome. This reasoning would suggest a continuance of the NPDES program for point sources while re-evaluating the current approach for storm water control, abandoned dump remediation and contaminated sediment removal.

Consensus Based Water Quality Improvements

Many in the regulatory agencies have verbally supported the end of command and control but have missed an obvious extension of this premise: consensus based water quality programs are largely voluntary. This is a very foreign concept to most regulators. Stated simply, command and control is administratively seductive. Often programs which begin as consensus based projects degrade into the old "comply or die" means of problem solving.

It is precisely this tendency which often prevents local units of governments (and most industries) from mitigating environmental problems on their own. Basically, it appears that most local units of governments don't trust state and federal regulators. Oddly enough, however, this is in direct conflict with the general public. Surveys of Rouge River residents confirm that the residents place extremely high trust in the Michigan Department of Environmental Quality (MDEQ) and the EPA and substantially less trust in their locally elected officials. Any consensus based water quality improvement program must recognize not only this incongruity but also the forces driving those differences.

The local units of government are under increasing pressure to provide more services while lowering taxes. When faced with a decision between a fire truck or a water quality program, the fire truck wins ever time. In short, local officials are doing the best they can with what they have available to them. Until the general public demands additional local programs, few are likely to be initiated.

The general public, on the other hand, has generally looked to the state and federal governments to protect their environment. Much of this is driven by their belief that the problems are too large to be handled on the local level. They continue to believe that big business and often their own local governments must be held accountable for their discharges. The public, however, does not recognize that this battle, for the most part, has been won. The battle now has shifted to a large number of smaller problems which are best fought at the local level. The challenge, therefore is to get the local officials and their constituencies to recognize their respective roles and encourage them to step forward.

The Rouge Project has struggled with the conflicts inherent in attempting to implement programs on a watershed basis where the political boundaries do not match the watershed boundaries. To overcome this problem, a series of subwatersheds have been identified and individuals from within the local units of governments were asked to lead small working groups which included officials from their neighboring communities. While the Rouge Program Office (RPO) provided a detailed menu of possible projects, it is these local officials who must determine the actual components of their storm water management program. This has some inherent strengths and weaknesses. Obviously, the local officials have the best understanding of the problems facing the local communities. It is equally obvious that each subwatershed will have different programs to address the various problems. Thus, the larger challenge is to integrate these programs in an equitable manner.

Process

Under the auspices of the Rouge Project, the Wayne County Department of Environment (WCDOE) invited the communities and other stakeholders within the Rouge River watershed to participate with the Surface Water Quality Division of the MDEQ in drafting the elements of a new Watershed Based Storm Water General Permit. The RPO provided staff assistance to the Core Advisory Group representing Rouge community stakeholders.

The process was based on building a consensus among the local agencies on the cost effective steps that can be taken to progressively address identified water quality issues related to storm water. As a result of this effort, the MDEQ issued a general permit that encourages local cooperation in addressing water quality problems on a watershed basis and provides an alternative voluntary approach to meeting state and federal requirements.

Content and Methodology for Developing the General Permit

General permits have been used by both the MDEQ and the EPA to administer various provisions of the Clean Water Act and specifically storm water. General permits differ from individually issued permits since the provisions of a general permit are the same for all applicants and do not require individual determinations based on site specific conditions. Once the provisions of a general permit are adopted by the regulatory agency, those who meet the criteria simply document that they meet the criteria and, if they do, are notified by the regulatory agency that they are covered under its terms.

The following permit attributes and application requirements are required under the Michigan Watershed Based Storm Water General Application/Permit. Governmental entities that have ownership and control of separate storm water drainage systems are eligible for coverage under the General Permit. Industrial, commercial and governmental entities required to have storm water coverage under federal or state regulations are not eligible. The permit authorizes the discharge of storm water from the permittee's separate storm water drainage system, storm water commingled with discharges authorized under other NPDES permits and specific non-storm water discharges provided these discharges do not contribute to a violation of state Water Quality Standards (WQS.)

At the time a community requests coverage under the General Permit, they must submit some general information, an illicit discharge elimination plan and a public education plan. The following summariezes these requirements.

General Information: Identification of the applicant, the proposed watershed for which a watershed management plan will be developed, and information about the applicant's storm water drainage system must be submitted. Each known point source discharge of storm water for which coverage is requested must be identified.

Illicit Discharge Elimination Plan: An adequate plan to eliminate illicit discharges is required with the application. The plan must include an implementation schedule. MDEQ will review the plan, and if it is unacceptable, identify the areas that need to be addressed before permit coverage is granted. The plan shall include: 1) a description of a program to find, prioritize and eliminate illicit discharges and illicit connections identified during dry weather screening activities and 2) a description of a program to minimize infiltration of seepage from sanitary sewers and septic systems into the applicant's separate storm water drainage system.

Public Education Plan: An adequate plan for public education is required with the application. The plan must include an implementation schedule. MDEQ will review the plan, and if it is unacceptable, identify the areas that need to be addressed before permit coverage is granted.

Some specific items to consider in development of the plan are listed in the permit. The permittee should select and prioritize these and other actions to be accomplished considering site-specific concerns and the potential impact in its jurisdiction. Consideration should be given to coordinating aspects of the plan with other watershed jurisdictions.

Once a unit of government has received their certificate of coverage, a Watershed Management Plan and a Storm Water Pollution Prevention Program must be submitted within 2 and 2 ½ years respectifully. The following summarizes these requirements:

Watershed Management Plan: Watershed management provides a framework for integrated decision-making in order to: 1) assess the nature and status of the watershed ecosystem; 2) define short-term and long-term goals for the system; 3) determine objectives and actions needed to achieve selected goals; 4) assess both benefits and costs of each action; 5) implement desired actions; 6) evaluate the effects, actions and progress toward goals; and 7) re-evaluate goals and objectives as part of an iterative process.

The MDEQ General Permit lists the following minimum requirements to be included in the Watershed Management Plans:

- An assessment of the nature and status of the watershed ecosystem.
- A definition of the short-term goals for the watershed.
- A definition of the long-term goals for the watershed, which shall include protection of designated uses of the receiving waters as defined in Michigan's Water Quality Standards.
- A determination of the actions needed to achieve the selected short-term goals for the watershed.
- A determination of the actions needed to achieve the selected long-term goals for the watershed.
- An assessment of both the benefits and costs of each action.
- A commitment, identified by specific permittee or others as appropriate, to implement the actions necessary to initiate achievement of the selected long-term goals by a specified date.
- Methods for evaluation of progress, which may include chemical or biological indicators.

Storm Water Pollution Prevention Program (SWPPP): SWPPP shall be submitted to the District Supervisor for approval. This will likely include a modification of the Illicit Discharge Elimination Plan and the Public Education Plan. The SWPPP incorporates the permittee's specific actions identified in the Watershed Management Plan and other actions the permittee will undertake to reduce the discharge of pollutants in storm water to the maximum extent practicable. The SWPPP also includes:

- A description of methods of assessing progress in storm water pollution prevention.
- A schedule for submittal of annual reports and minimum requirements for the annual reports.
- The identification of a designated contact person.
- The retention of records.

Conclusion

Much work has been performed to establish the concept of a General Permit for watersheds. The program has been enbraced by the federal government, the state government and the local units of government. The success or failure of this program can only be measured by monitoring the in-stream water quality. Small positive changes can already be seen in the Rouge River. As the project proceeds, the cummulative impact of a lot of little actions will be realized. This process provides for an adaptive management feedback loop. The local units of government are free to pursue what works and abandon what doesn't. Most importantly, progress will be made and be made in a timely manner. The General Permit process in Michigan can document that the watershed approach is the appropriate way to proceed and not simply an excuse to delay.

SOUTH PLATTE

Moderator, Ben R. Urbonas

Maintaining an Urban River

Bryan W. Kohlenberg, P.E.[1] and Ben R. Urbonas, P.E., M. ASCE [2]

Abstract

Since 1987 over 100 restorative and cooperative projects have been completed along the South Platte River in the metropolitan Denver area. This paper describes the use of routine and restorative maintenance activities by the Urban Drainage and Flood Control District to clean up, stabilize, and rehabilitate riverbanks, the river's longitudinal grade, structural facilities, and vegetation. This paper also overviews the continual coordination with federal, state, and local governments and regulatory and planning personnel.

Introduction

The Urban Drainage and Flood Control District (District) was established by the Colorado State Legislature in 1969 for the purpose of assisting local governments with multi-jurisdictional drainage and flood control problems in the Denver metropolitan area. In 1985, the District cooperated with several local governments to develop a Major Drainageway Plan (MDP) for the 65.6 kilometers (41-mi.) reach of the South Platte within District boundaries. The MDP identified problems related to riverbank erosion and riverbed degradation and recommended how to address these problems. It was evident from the MDP study that, like most rivers in an urban area, the South Platte has been used, abused and often neglected. Several years ago the river reach through Denver was characterized as a "sad, bewildered, nothing of a river" – a man-made ditch.

In response to the MDP recommendations, the South Platte River Program was established in 1987. Its goal is to address the unique problems associated with a

[1]Project Manager, South Platte River Program,
[2]Chief, Master Planning & South Platte River Programs,
Urban Drainage & Flood Control District, 2480 W. 26th Avenue, #156B, Denver, CO, 80211

multi-jurisdictional, dynamic, fluvial waterway in an urban area. The program's objectives are as follows:

- Maintain and restore existing flood carrying capacity.
- Protect properties and facilities adjacent to the river from damages.
- Arrest the river's vertical degradation and excessive lateral migration so often associated with riparian area destruction.
- Work with local governments, landowners and others to implement the MDP.
- Acquire right-of-way for flowage and maintenance access to the river.

The District's Board of Directors annually allocates maintenance funds for this program on the basis of relative need, priorities of local jurisdictions and compatibility with the South Platte River MDP. The annual maintenance and cooperative projects budget varies year to year between $500,000 and $1,200,000. Maintenance activities requested by private property owners (i.e. cooperative projects) require a minimum contribution of 25 percent from the property owner. In most cases, the dedication to the District by the property owner of a permanent easement satisfies the property owner's share of the project.

Routine and Restorative Maintenance Activities

Each year the District hires local contractors to routinely remove trash and debris up to five times along 30 miles of the river. Each removal generates an average of 18 truckloads of trash and debris taken to a landfill. Local government personnel or volunteer groups pick up additional trash. Trash is also removed from trash receptacles maintained by park personnel along recreational trails. The District also contracts for mowing and tree trimming along the public trails. This ensures the trails remain clear of vegetation and are accessible to the public. Although this type of routine maintenance often is not noticed by the public, without it the South Platte River corridor would have an entirely different "look."

Restoration maintenance is also contracted for. It primarily consists of the following types of projects and typical tasks:

Projects

- River Bank Stabilization and Restoration
- River Grade Stabilization
- Existing Structure Restoration
- Trail Preservation
- Utility Crossing Protection

Tasks

- Project Identification
- Design Surveys and Mapping
- Engineering, Landscape and Ecological Design
- Right-of-Way Acquisition
- Project Permit Acquisitions
- Construction and Oversight

Detailed inventories of facilities and properties along the river are maintained to assist maintenance operations. Annual field surveys of 53 river cross-sections at fixed locations help the District track and assess horizontal and vertical movement of the river. They provide "early warning" of impending problems.

Riverbank stabilization and restoration, river grade stabilization, and the restoration of deteriorating structures tend to dominate restoration projects. Figure 1 shows a typical engineering cross-section used to stabilize a badly eroded bank. Although riprap is used and needed (without it the vegetated slopes become undermined and banks tend to erode again), it is fully mixed with soil, buried and revegetated with native grasses, shrubs and willow.

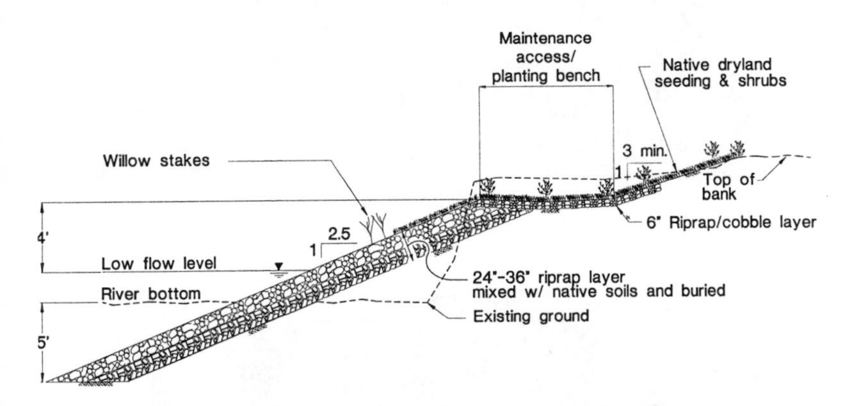

Figure 1. Typical Cross-Section of Bank Stabilization/Restoration

Figure 2 shows how a badly eroded riverbank was restored. Here the vertical bank was regraded, clean concrete rubble imported, broken up and placed into the bank below a layer of buried rock riprap, and then the entire site was revegetated. The site is now covered with native grasses, willows, shrubs and cottonwood trees.

Figure 3 shows the results of an exposed sewer line rock-buttressing project below Chatfield Reservoir Dam near Denver. The river bottom had severely down-cut due to water being detained and released "sediment hungry" from the upstream dam. A six-foot drop developed below the once buried sewer line. This situation made safe boat passage impossible and eliminated any chance for fish migration. With the help of the local governments and the utility company a sloping grouted boulder drop with combination boat/fish passage chute was installed in 1996.

Figure 2. A during and after example of bank restoration

Figure 3. Before and After Buttressing of Existing Sewer Crossing

General 404 Permit

In 1987, after almost two years of negotiations with the U. S. Environmental Protection Agency (EPA) and the U. S. Army Corps of Engineers (COE), the District was issued a general restorative maintenance 404 permit. This permit specifically allows for, within clearly defined quantity and construction method limitations, certain restorative maintenance activities and structures such as bank stabilization, utility structure buttressing, fallen tree bank revetments, certain types of jetties and rock grade control structures up to three feet in height. This permit was renewed for a second time in 1997.

Prior to the issuance of the General Permit, most restorative maintenance projects had to obtain an Individual 404 permit, which took between three months and two years to obtain. With the General Permit, individual maintenance project permits now take less than one month to obtain. As a result, the District is now able to respond to restorative maintenance needs quickly and at much less cost than previously. The General Permit, however, does not exempt the District from endangered species surveys or other special circumstances, such as preservation of discovered archeological sites.

Operating a Maintenance Program

The following key suggestions are offered for how to best operate and manage an urban river maintenance program:

- Create one river maintenance budget to be used at agencies discretion;
- Hire pre-qualified contractors (if possible) familiar with work in rivers;
- Need to provide on-site construction guidance due to dynamic nature of river channel;
- Realize that "bioengineering" streambank stabilization techniques are actually inappropriate for many urban streams and rivers;
- Work with COE and local governments to expedite permits for maintenance activities.

Conclusion

Maintenance programs can be used effectively and economically to restore degraded and abused riverine environments. They need to operate, however, under the COE's Section 404 permit program and the local government's floodplain regulations. The cost of obtaining and administering individual permits for a large number of maintenance projects can be unwieldy, time consuming and costly. A General Permit, similar to the one that has served the South Platte River Denver area, can help expedite each project. Also, local governments need to try to establish a working relationship with the permits branch of the COE. It also means that Federal agencies such as U.S. Fish and Wildlife Service, COE, EPA, and/or their State counterparts need to be convinced to have a level of trust in the local agencies doing the work.

Mother Nature can be severe in riverine environments and can blow our best intentions "out of the water" even when we try our hardest. Eventually even small, pay-as-you-go, maintenance projects can add up to significant riverine restoration.

South Platte River Enhancement - North Denver

David J. Love, P.E.[1]

Abstract

This $11,000,000 river enhancement project is located at the northern edge of the City of Denver in the Globeville neighborhood. This high profile project, driven by the Mayor's Office and the Urban Drainage and Flood Control District, constructed a greenway trail, aquatic and terrestrial habitats, reclaimed and stabilized the river banks, constructed flood levees and floodwalls to provide 100-year flood protection, enhanced the wetland and riparian areas along the river, and provided educational signage throughout the project reach. This paper describes the project and provides insight into the need for public participation.

Introduction

The purpose of this project was to design and construct flood control and riverine improvements within approximately a one and one half (1.5) mile reach of the South Platte River within the Globeville neighborhood. This project is the largest single flood control improvement project undertaken by the City of Denver. The South Platte River within the Globeville neighborhood typically is narrow with high and steep banks. Before this project began, the public and natural amenities located within this reach of the river included: an 8-feet wide bicycle/pedestrian trail along two-thirds of the reach; an abandoned sewage treatment plant; non-native riparian vegetation; poor quality wetlands; a small pocket park; and a small community park.

Over time, man has narrowed and straightened the South Platte River, filled its banks, and constructed bridges and irrigation diversion dams which have affected its flood conveyance capacity. Under existing conditions, the 100-year flood will overtop the banks along several portions of this reach of river. The floodwaters, which overtop the west bank, cannot readily rejoin the floodwaters within the

[1] President, Love & Associates, Inc., 2995 Centergreen Court South, Suite C, Boulder, Colorado 80301-5421

river because the 100-year floodplain area west of the river is lower than the west bank of the river's channel. As a result, approximately 300 acres of land remained within the floodplain boundaries.

This project addresses not only the flood protection needs of the community but also provides for significant restoration of the river and its adjacent riparian habitat. Public input into the design of this project has been extensive during many work sessions, field tours, and public meetings.

Project Overview

Development over the last 100 years along the study reach has reduced riparian habitat, diminished the diversity of fish and wildlife species, steepened channel side slopes, and modified the properties of the alluvial soils. The aesthetic and recreation amenities along the study reach are also lacking. River improvements within the study reach are severely constrained by minimal rights-of-way, accessibility, bridges, diversions, river crossings, existing utilities, and adjacent land uses.

The banks of the river within the subject reach have been filled by man over time. The fill consists of sandy, gravelly, and/or clayey material and can include construction debris and trash. The reach currently has only a limited habitat for fish. Fish currently congregate near bridge piers, in backwater areas, and under overhanging vegetative cover. Wildlife is reported to include geese, ducks and other waterfowl, song birds, rats, feral cats and other small mammals, beaver, and an occasional deer.

Trees, shrubs, grasses and weeds within the river corridor are primarily non-native species that do not provide optimum habitat for wildlife. Four plant communities exist along this reach of the river, including point bars, herbaceous wetlands, shrub wetlands and riparian forest. It has been determined by the U.S. Army Corps of Engineers that the proposed wetland and habitat opportunities created by this project far outweigh any disturbance that will result from the project.

At the downstream end of the study reach, a concrete diversion dam diverts water to the headgates at the entrance to the Burlington Ditch and O'Brian Canal. The dam extends entirely across the main channel of the river and can divert approximately 1,000 cubic feet per second (cfs) through three existing diversion gates. At times, the diversions are capable of drying up the South Platte River downstream to Sand Creek.

Opportunities and Constraints

Participants from a wide range of government agencies, organizations, and individuals were active players of a working design committee which attended biweekly progress meetings during the design and construction phases of this project during the past three years. Multiple public "Town Hall" meetings were held with both residents and business owners in the Globeville neighborhood in order to solicit input as to desires and expectations that would result from the completion of this project. Key results requested by the majority of the neighborhood participants included removal of the 100-year floodplain from their homes and businesses, increased safety, creation of aquatic and wildlife habitats, and creation of jobs along the corridor due to the enhancement of the river corridor and to the removal of properties from the floodplain. This reach of the South Platte River was inhabited by numerous "homeless" persons, was fairly isolated and was used as a trash dump site on a daily basis. This area was sorely in need of improvement in order to provide a basis for the revitalization of the neighborhood which was so necessary to its economic viability.

This project affords many opportunities for benefits to the community. Some of these include reduction of flood damage and life safety hazards to existing properties; restoration of a riparian corridor within an urban setting; boat chutes with fish ladders; creation of recreational uses, in-stream and trail; improved fish and wildlife habitat; increase in value of adjoining properties; improved corridor image and aesthetics; added park opportunities; added maintenance access; and public awareness and benefit.

Some of the constraints that have had an impact on the project are limited right-of-way, in-situ soils quality, existing structures and river crossings, existing utilities, irrigation diversion structures, adjacent land uses, existing limited quality vegetation, accessibility from neighborhoods and adjacent properties, financial resources, dwindling populations of fish and wildlife, increased maintenance cost, and beaver control for vegetation.

Major objectives within this study reach are to eliminate the overtopping of the Franklin Street Bridge during the 100-year flood, to prevent flooding along the west bank of the river, to construct a pedestrian/maintenance trail underpass and to install boat chutes/fish ladders for recreation and fish passage in this reach. To meet these objectives, the conveyance of the river upstream and downstream of the bridge will be increased by lowering the river invert in the vicinity of the bridge, which requires the construction of an upstream approach channel for the ditch headgates.

The headgate approach channel is designed to capture approximately the first 1,000 cfs of flow in the river. The mouth of the approach channel consists of a boulder drop weir about 200 feet in length. Water diverted over the boulder drop weir falls into the headgate approach channel.

After the existing diversion dam is removed and the approach channel and new diversion dam constructed, the main channel will be excavated to increase the conveyance capacity of the river. To control the grade of the river and provide recreational activities, boat chute weirs with fish ladders will be installed. The top of the downstream boat chute weir will be at grade to accommodate possible channel excavation downstream of the study reach in the future.

Levees and floodwalls will be installed on the west bank at select locations to contain the 100-year flood within the main channel. Floodwalls will be used only when constrained by limited Right-of-Way. Levees and floodwalls will be constructed to meet FEMA guidelines. These requirements include but are not limited to; adequate freeboard, closure devices on all openings, embankment protection against erosion, and assurance of embankment and foundation integrity through stability and settlement analyses. Cross Section No. 1 below shows a typical view of the river improvements when completed.

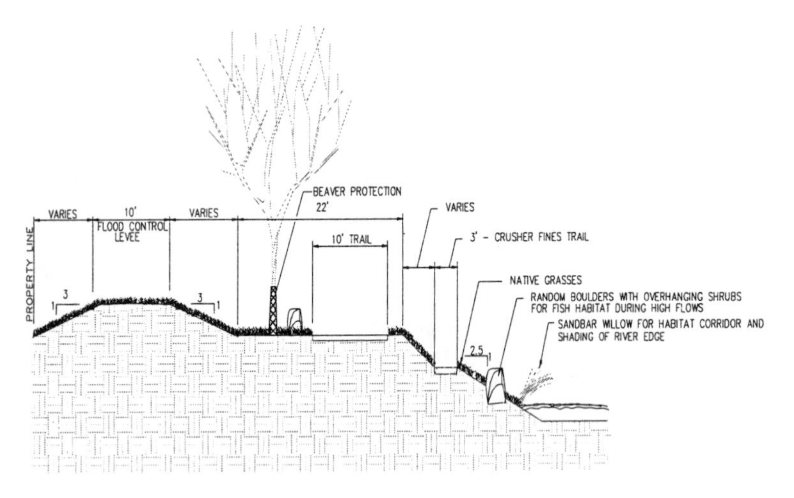

CROSS SECTION NO. 1

N.T.S.

Recreational amenities proposed for this project can be separated into two categories, instream recreational features and bank recreational facilities. Instream recreational features for this project include the installation of a combination boat chute/fish ladder structure. This will be located near the Franklin Street Bridge. This structure not only provides opportunity for boating enthusiasts but will also provide hiding places and migration opportunities for fish within the river. In conjunction with the proposed boat chute/fish ladders, hard surfaced boat landings are proposed at the upstream and downstream ends of the proposed boat chutes. Strategically placed boulders, physical barriers and directional/warning signs will be installed near the proposed boulder drop weir for the headgate approach channel for boater safety.

The construction of vegetated and non-vegetated jetties and flow deflectors will occur at various locations throughout the project reach to increase localized velocities, create scour holes, provide oxygenation within the main channel and to protect and create gravel point bars. Jetties on the west side of the river will be accessible from a soft surface trail for potential river enthusiasts, anglers and weekend aquatic biologists.

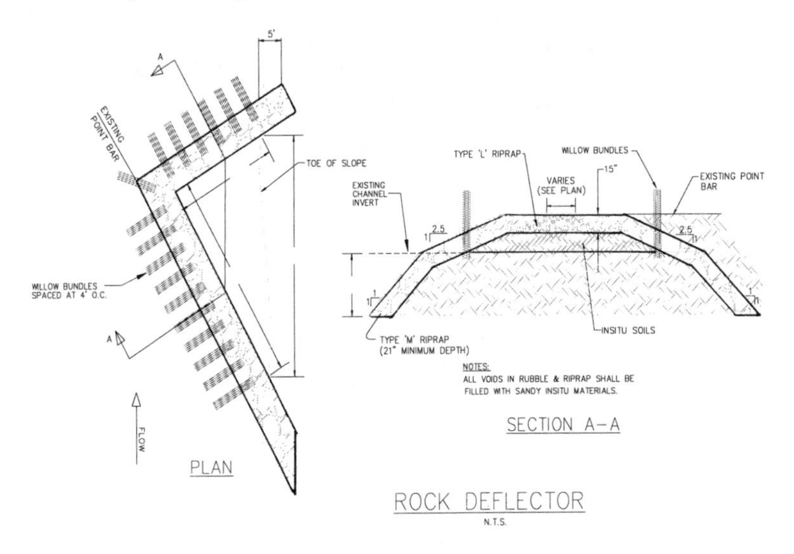

Bank recreational amenities include the construction of a 10-feet wide concrete pedestrian/maintenance trail along the westerly bank of the river adjacent to the Northside Treatment Plant. The proposed improvements will complete trail connections to the adjacent Adams County trail at the downstream end of the project and the existing Denver Greenway Pedestrian trail. A mile reach of 8-feet wide concrete trail will be widened to 10 feet with 2-feet wide shoulders added for safety.

The project reach is currently vegetated with native and non-native grasses and select areas of Siberian Elm and Plains Cottonwood which are generally located at the ordinary high water line. The toe of the channel bank is severely eroded and devoid of vegetation. Proposed improvements include willow live staking along the toe of the bank and around the edges of the proposed jetties. The willow, when mature, will provide shading of the river edge which should lower water temperatures, provide hiding zones for aquatic communities, and create a linear corridor for wildlife migration. Randomly placed boulders will be interspersed along the bank and within the river to create additional fish refuge zones. Shrubs will be planted at the ordinary high water level to insure that low velocity zones and hiding places for fish will be available during times of higher flow depths.

Live staking of cottonwoods, at or near the ordinary high water level, will occur at random locations along the project reach to provide additional shading and habitat for wildlife along the banks of the river. Below a detail used has been provided. More formal landscaping consisting of native trees, shrubs, and grasses and wildflowers will be installed adjacent to the trail. All trees will have screening to protect them from beavers. Large boulders will be situated in groupings to provide resting and observation opportunities.

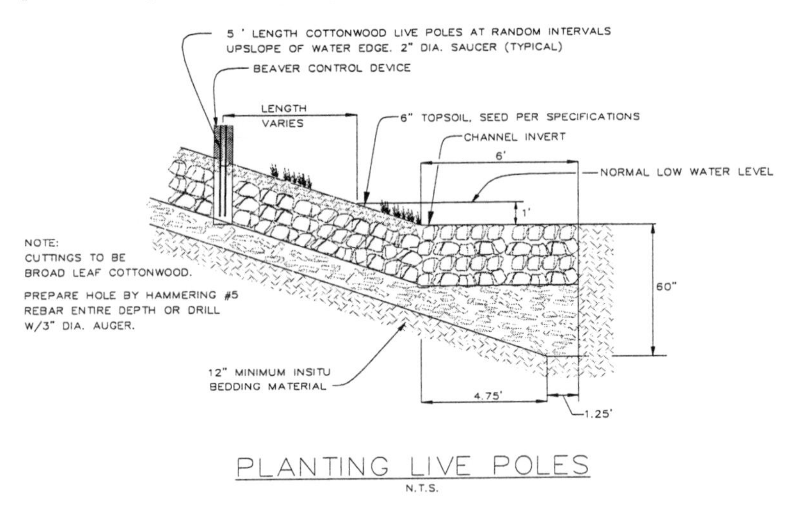

PLANTING LIVE POLES
N.T.S.

Conclusion

The goals that were set at the inception of this project will not only be met but will be exceeded once all three phases of the project have been constructed. The final missing link of the Platte River Greenway trail has been completed; approximately 300 acres of land will be removed from the 100-year floodplain which will allow for a sorely needed revitalization of the Globeville community; the ditch company will be able to deliver its water rights to its users as it historically has done; boaters will be able to navigate the river safely with the removal of the diversion dam and construction of multiple boat chute drops; fish passage can now occur with removal of the dam and the aquatic life will find new hiding places and enjoy the shady coolness of the waters from the vegetation planted along the river's banks and the wildlife and terrestrial habitats should prove inviting to a wider array of species. This reach of the South Platte River will now allow for greater safety to users, prosperity to the land owners and businesses and a more enjoyable recreational experience for users.

Denver's South Platte River Corridor Project

Marc Alston, P.E.[1]

ABSTRACT

In 1995, Denver Mayor Wellington Webb endorsed a set of recommendations for enhancing Denver's 10.5 mile South Platte River Corridor, and created the South Platte River Commission. The recommendations, contained in "A Vision for the South Platte River", covered improvements to multiple uses of the river: recreation, boating, fishing, open space, wildlife habitat, education, youth employment, community awareness and flood control. This paper describes the wide variety of coordinated projects that have followed those recommendations. These projects include over one hundred acres of park expansion, a managed flow program, channel improvements, riparian restoration, river clubs in the schools, volunteer revegetation, youth crews and river cleanups and celebrations. The paper also overviews the South Platte River Commission, which is the basis for a thriving partnership representing city, state and federal agencies, special districts, and non-profit and educational organizations.

INTRODUCTION

The South Platte River flows for 10.5 miles through Denver, the center of a rapidly growing metropolitan area with a semi-arid climate. Overall, the South Platte runs for 400 miles from the Central Rockies to its confluence with the North Platte River in Nebraska. It is essential for many uses in Colorado: agricultural, water supply, waste disposal, wildlife, and as a recreational amenity.

While the South Platte is a major stream in Colorado, flows are low when compared to other parts of the U.S. Flows vary from less than 100 cfs in the fall and winter dry periods, to 2,000 cfs of average maximum runoff in June. Flow in the river is heavily managed for flood control, water rights and water supply purposes.

[1] Coordinator, South Platte River Corridor Project, City and County of Denver Mayor's Office Room 350, 1437 Bannock Street, Denver CO 80202 (On loan from EPA Region VIII)

HISTORY

In *Centennial,* James Michener's novel about the settlement of Colorado's high plains, the South Platte was described.

" And finally there is the river, a sad, bewildered nothing of a river. It carries no great amount of water, and when it has some it is uncertain where it wants to take it. No ship can navigate it, nor even a canoe with reasonable assurance. It is the butt of more jokes than any other river on earth, and the greatest joke is to call it a river at all. It's a sand bottom, a wandering afterthought, a useless irritation, a frustration, and when you've said all that, it suddenly rises up, spreads out to a mile wide, engulfs your crops and lays waste to your farms.

Its name is as flat as its appearance, the South Platte, yet for awhile it was the highway of empire. It was the course of stirring adventure and the means whereby the adventurers lived. Once mighty enough to help build a continent, it is now a mean, pestiferous bother." (Michener, 1974)

Denver's origins are tied to the banks of the South Platte due to the discovery of gold in the 1850s. The River was very different before 2,000,000 residents surrounded its banks. The South Platte was much wider, not confined to a defined channel, and subject to major yearly flooding. Unfortunately, as Denver grew, its River quickly became a place in which you put your waste and turned your back.

The South Platte flows for 10.5 miles through present day Denver.

In the 1970s attitudes about urban rivers began to change. Joe Shoemaker, at the request of Mayor Bill McNichols, set up the Greenway Foundation, which played the key leadership role in looking at the South Platte River through Denver as an amenity to reclaim, nurture and protect. Between 1975 and 1983, the Platte River Greenway was established and served as a national model for urban river restoration. The term Greenway was coined in Denver.

Between the mid 1980s and early 90s, efforts to improve Denver's river lost momentum. In late 1994, Mayor Wellington Webb formed the South Platte River Working Group to explore opportunities for enhancing the River as it passes through the City and County of Denver. The Working Group was composed of numerous river interests representing city, regional, state, and federal agencies, as well as private and non-profit stakeholders.

In February, 1995 Mayor Webb endorsed the Working Group recommendations which were entitled "A Vision for the South Platte River". The Denver Post headline of February 3, 1995 was "Webb river plan hailed as miracle". Recommendations included :

* insuring flow and quality levels needed to support recreational and environmental needs along the river, including fishing and boating;
* improvements for riparian aquatic habitat and flood control;
* creation of additional parks
* enhancing care and safety;
* an added focus on environmental education;
* increasing community awareness of the River as a valuable natural resource. (Webb, 1995)

Among the recommendations was creation of the South Platte River Commission, which was established under Executive Order on February 2, 1995. The responsibilities of the 28 member Commission are to:

* make recommendations on needed river improvements and opportunities;
* help create partnerships among governmental, private and non-profit entities;
* plan and pursue funds for improvements;
* share information on River activities;
* help coordinate River activities with other jurisdictions;
* oversee implementation of "A Vision for the South Platte River.

THE PRESENT: DENVER TAKES OWNERSHIP OF ITS RIVER

While the challenges facing Denver in addressing these objectives were substantial and somewhat intimidating, an aggressive and positive approach has resulted in huge progress. Much of the success to date is due to the partnering flowing from the South Platte River Commission. A broad collaborative approach has been taken to improving the multiple uses of the river: fishing, boating, recreation, wildlife, open space, education and flood control. Major undertakings have resulted in successes that apply to the entire 10.5 miles of river in Denver and also to specific, individual stretches. This was recognized nationally when the South Platte River Commission received the American Rivers Gold Medal for Urban River Partnership in 1997.

Obtaining funding support and leveraging of funding opportunities has also been a key factor in the headway made to date. The City and County of Denver has contributed over $4M per year over the past three years. Urban Drainage and Flood Control has made yearly contributions of over $1M. A Legacy Grant of $6M was awarded from Great Outdoors Colorado (state lottery funds) in 1996. Several other federal and state sources have been used to finance improvements. These funds have focused on a broad assortment of exciting projects.

PROJECTS: PARKS, RIVER RESTORATION AND WATER MANAGEMENT

Denver's Greenway provides a natural linkage with its park system. Several riverside parks were integrated into the original Greenway system. Current efforts will result in major park additions adjacent to the river in south, central, and north Denver.

⇒ In southern Denver, *Grant-Frontier Park* will be expanded and directly connected to the river by closing a section of road that separates the park from the Greenway trail and river. This will greatly expand riparian wildlife habitat, improve safety, and reconnect the neighborhood to the river. These improvements will be complete in early 1998.

⇒ The *Riverfront Park System* is the crown jewel of the 10.5 miles of river corridor through Denver. Since 1996, this 1.5 mile stretch encompassing 6 parks (*Gates-Crescent, Fishback, Centennial, Commons and City of Cuernavaca*) has seen rapid improvement. Riverfront will ultimately include over 60 acres of new park space, all of which is expected to be completed by the end of 1999.

⇒ The thirty acre land acquisition for *Commons Park* was a major hurdle for the City. With the completion of Commons, the vision of a large park in the Central Platte Valley, first envisioned in 1893, will become a reality.

Rafting through Riverfront Park

⇒ One mile of River channel adjacent to Commons Park will see enhancements specifically designed to improve the fishery and boating, with completion in 1999.

⇒ At Denver's northern border, the city's abandoned wastewater treatment facility is undergoing redevelopment. The 70 acre *Northside Treatment Plant (NTP)* will contain a 13 acre park, a National Guard Armory, a wildlife refuge, and 23 acres of industrial park. Demolition of treatment plant facilities will be complete by fall 1998 to allow for park construction in 1999. The wildlife refuge has some of the best wildlife value in Denver.

⇒ The *North Denver River Restoration Project* will improve 1.3 miles of channel, provide flood control for the neighborhood, restore native vegetation, and complete the final link in Denver's Greenway. This project is adjacent to the NTP.

⇒ Additional recreational improvements will result from tributary trail replacements and a new tributary Greenway along Sand Creek.

Due to multiple demands for water, the South Platte is heavily managed, and over-appropriated at many times of the year. Flows are often too low for boating, and for an optimal fishery. Goals have been set by the SPRC that call for establishing a warm water fishery and enhanced boating.

⇒ Under an innovative approach by the Denver Water Department, a managed flow program for warm weather months attempts to meet a target flow of 150 cubic feet per second. This cooperative venture, unique in Colorado, has been initiated to enhance boating and the fishery. The Water Department is planning further efforts to ensure that these flow goals can be met for the long term.

⇒ Water quality is degraded from non-point sources, stormwater, wastewater discharges and historical urban use. Water quality impacts the fishery and recreational uses. Swimming in the river is not a designated use. Denver is developing options to improve water quality to benefit the fishery and recreational uses. Improving the fishery will take a long term approach that is likely to have high costs and institutional difficulties.

Youth do riverside trail work

YOUTH AND EDUCATION

Use of the River as an educational resource and a source of jobs for youth has grown rapidly through the efforts of the SPRC. Platte River Clubs have been created in 20 Denver middle schools, including 2 alternative schools. Five hundred students participated in the clubs. The cooperative program with Denver Public Schools is expanding to both the elementary and high schools during the 1997-98 school year.

The South Platte River after school clubs offer a variety of multi-disciplinary academic and recreational activities in addition to learning about the historic and environmental value of the river and riparian area. Club members also learn about future career opportunities related to the care and enhancement of natural areas. Activities include study of wildlife and plant species, water quality testing, assisting in maintenance of trails and parks, and rafting excursions. The Club program earned Denver the U.S. Conference of Mayors 1997 City Livability Award.

Summer employment opportunities also are growing. The River Guide and River Ranger programs provide unique work experiences; result in hundreds of raft trips taking place during warm weather; result in hundreds of man hours focused on the enhancement and protection of the river corridor; and provide information and assistance directly to public visitors to the Greenway.

COMMUNITY AWARENESS AND PARTICIPATION

There have been enormous increases in awareness and visibility for the River, with increases in neighborhood participation and volunteering. On

January 1, 1996, Mayor Webb proclaimed 1996 "The Year of the South Platte River" in Denver. Visibility was given to groundbreakings, building demolitions, land acquisitions and youth projects through press releases and other announcements. Hats Off to the Platte was held as a public celebration of the River in 1995 and 1996. Vice President Gore visited the River. A major cleanup day was held in the Spring. Mayor Webb hosted over ten press conferences and events along the River during 1996.

These types of activities continued in 1997 with the first River Rendezvous, which celebrates the history of the River as Denver's birthplace. Volunteers for Outdoor Colorado assisted Denver by organizing a Revegetation Day for 1,000 volunteers. A yearend celebration for the South Platte River Clubs attracted over 400 people. Planning is underway for 1998 for River Rendezvous, the South Platte Club celebration, and a Spring cleanup to expand into neighboring counties.

Volunteers for Outdoor Colorado Day

Neighborhood coordination is an important aspect of project planning. Local meetings are held to gain input, and to keep people aware of project progress.

THE SOUTH PLATTE RIVER AS A DENVER PRIORITY

Commitment to the river has become a solid and continuing reality since 1994. Tremendous momentum and major support is now in place to establish Denver's vision of the River as an amenity to be cherished by its citizens. Under Mayor Webb, Denver will continue to move forward aggressively to meet the goals of the 1995 Vision Report: " If we protect and care for our River and help restore its beauty, the South Platte will bring our children and theirs unmatched recreational, educational and development opportunities." (Webb, 1995)

The significance of the South Platte as it relates to both Denver's past and future is monumental. There will be continued challenges, and hopefully continued success, in maximizing the River's value to the community for many years to come.

APPENDIX

REFERENCES

1. Michener, James A., Centennial, Random House Inc., 1086 pg., 1974.

2. Webb, Mayor Wellington E., and the South Platte Working Group, Imperative 2000: A Vision for the South Platte River, 27 pg., 1995.

WETLANDS

Moderator, Ed Herricks

"Consensus Based Planning: Developing a Conceptual Plan for Everglades Restoration"
Stuart J. Appelbaum
Corps of Engineers, Jacksonville District

"Wetlands Ecology and Water Resource Management"
Steven I. Apfelbaum
President, Applied Ecological Services, Inc.

"Developing a Wetland Mitigation Bank – A Private Sector Perspective"*
John Ryan
President, Land and Water Resources, Inc.

"Hydrologic Effects of Wetlands in Southeastern Wisconsin"*
D. L. Hey and Jeffery Wickenkamp
Hey and Associates

*Text not available at time of printing.

Consensus Based Planning: Developing a Conceptual Plan for Everglades Restoration

Stuart J. Appelbaum, M. ASCE[1]

Abstract

The Central and Southern Florida (C&SF) Project is a multi-purpose water resources project which was built by the Corps of Engineers (Corps) to provide flood control; water supply for municipal, industrial, and agricultural uses; prevention of saltwater intrusion; water supply for Everglades National Park; protection of fish and wildlife resources; and other services to the south Florida area. While the project has served its authorized purposes well, it has also had unintended adverse consequences on the unique Everglades and Florida Bay ecosystems. In 1997, Congress authorized the Corps to undertake a comprehensive review of the C&SF Project to determine the feasibility of modifying the project to restore the south Florida ecosystem while meeting other water-related needs. In June 1993, the Corps began the C&SF Project Comprehensive Review Study (Restudy). The Restudy is being accomplished by an interdisciplinary, multi-agency planning team led by the Corps and its cost-sharing partner, the South Florida Water Management District.

In March 1994, Governor Lawton Chiles established the Governor's Commission for a Sustainable South Florida to recommend strategies for ensuring the long-term compatibility of a strong south Florida economy and a

---

[1]Chief, Ecosystem Restoration Section, U.S. Army Corps of Engineers, Jacksonville District.

healthy south Florida ecosystem. This 47 member commission consists of business, agriculture, government, public interest, and environmental organization representatives. In October 1995, the Commission completed its *Initial Report* which contained 100 recommendations for achieving sustainability in south Florida.

Acting on one of its initial recommendations to provide specific recommendations describing its preferred alternatives for the Restudy, the Commission, with the assistance of the Restudy team, developed planning objectives, selected a list of 40 preferred options to be evaluated, and incorporated these options into a *Conceptual Plan for the Restudy* which was completed in August 1996. The Commission's Conceptual Plan for the Restudy includes ongoing water resource projects grouped into 13 thematic concepts with additional elements, which collectively will help achieve a sustainable South Florida. The conceptual plan was developed using a facilitated process to achieve consensus of the Commission members. This process was aided by formulation workshops conducted by the Restudy team. In accordance with the Water Resources Development Act of 1996, the conceptual plan will be used as a starting point by the Restudy team for the development of a comprehensive plan that will be prepared by July 1, 1999.

This paper will discuss the process used to develop the conceptual plan. The paper will focus on discussion of the role of the Restudy team in developing and presenting technical information used by the Governor's Commission in the development of the conceptual plan.

WETLAND ECOLOGY AND WATER RESOURCE MANAGEMENT

Steven I. Apfelbaum[1]
J. Douglas Eppich[1]
Timothy R. Pollowy[1]

Abstract

Studies of the Des Plaines River provide evidence that many existing streams did not have conspicuous channels and were not identified during presettlement times (prior to 1830s in the Midwestern U.S.). Many currently identified first-, second-, and third-order streams were identified as vegetated swales, wetlands, wet prairies, and swamps in the original land survey records of the U. S. General Land office. A review of historic data indicates increases in discharge during low, median, and high flows since settlement. The modern channels formed inadvertently or were created to drain land for development and agricultural uses. Current discharges may be 200- to 400-times greater than historic levels, based on data from 1888 to the present for the Des Plaines River in Illinois (Apfelbaum, 1997).

Land development has resulted in a change from diffuse overland flows to increased runoff, concentrated flows, and significantly reduced lag time. The opportunity to emulate historic stormwater behavior exists through integration of natural ecosystems in urban and agricultural landscapes. Solutions that are easier to maintain, less expensive, more attractive, and offer other benefits as compared to many conventional stormwater management solutions are now available.

---

[1]Applied Ecological Services, Inc., 17921 Smith Road, P.O. Box 256, Brodhead, Wisconsin 53520-0256

Introduction

Both now and historically, wetlands have played an important role in maintenance of regional water balances. They also contribute to other very important local, regional, national and global levels of performance of stormwater and floodwater management.

Historic evidence from major streams and river systems in the central United States suggests that they have changed substantially. It is important to understand the magnitude of these changes in rivers, wetlands, and their tributary upland ecosystems to comprehend the change in hydrology and hydraulics these systems have undergone. Future engineering approaches to stormwater management, even in highly urbanized environments with limited open space, could benefit greatly utilizing restored ecological systems (e.g. prairies, wetlands, forests, etc.) for creative and cost effective solutions. These ecological systems perhaps better address flood management and stormwater, and provide a series of secondary benefits (e.g. increased wildlife habitat, increased biodiversity, water quality enhancement, and additional open space) that may not be provided by conventional approaches to stormwater management design.

Historic Hydrologic System Functions

If rivers are indicators of watershed change, then a review of how the hydrology of rivers has changed should be useful. This understanding indicates the magnitude of the changes that may be addressed by thinking about wetlands, prairies, and other landscape elements of ecological systems as functional levels in water resources management.

Studies of the Des Plaines River have identified regional (watershed) hydraulic and hydrologic changes associated with development from initial land clearing associated with farming to present day urban development. Over 90% of the historic native vegetation in the Des Plaines River watershed including wetland, prairie, savanna and forest systems has been lost or severely degraded. This has resulted in increased water runoff and sediment loads; lessened stream geometry stability; decreased stream system functions; deteriorated water quality; lessened river habitat; and decreasing human quality of life opportunities.

What opportunities exist in urban and rural areas for reestablishing some percentage of the historic wetlands, prairies and forests? What benefits might be realized? What are the associated costs? Are there other benefits besides water quality and flood management that might offset potential costs? These questions are all fundamental to ask when trying to understand the potential opportunities for incorporating natural systems and water resources. A series of residential and commercial projects have been constructed that have begun to answer these questions. Owners of these projects have decided that the benefits of restored prairies and wetlands are greater than the costs.

The Prairie Crossing Project

Typical urban development causes increased stormwater runoff rates and volumes, and an increased runoff of contaminants associated with developed land use. Contaminants include sediments, heavy metals, fertilizers, de-icing materials, and many other chemical constituents. Typical residential developments maximize building density and incorporate open space into individual lots. Public open space is only provided where required by ordinance for use as parks or for stormwater management purposes. Stormwater management for urban development is typically concerned only with minimizing onsite and downstream flooding, and other nuisance aspects of stormwater runoff. Consequently, most urban stormwater systems consist of storm sewers to convey runoff, a detention basin, and an outlet structure to control stormwater release rates.

The Prairie Crossing project, a large residential development in Lake County, Illinois, has taken a series of measures to reduce stormwater runoff rates, volumes, and pollutant loading. These measures include integrating source controls and large-scale restored landscapes into the development as a major element of the stormwater management system. The stormwater management system consists of upland prairie biofiltration, natural open swale conveyance systems, wetlands, and a lake. In combination, these increase lag time, increase opportunities for pollutant removal through settling and biofiltration, and reduce the rate and volume of runoff through enhanced infiltration opportunities. Prior to development, the site was farmed under an annual crop rotation; soils were modified by drainage improvements, including an extensive tile system, and the native biological communities were eliminated.

The Prairie Crossing project includes a high-density "village center" and an outer area of cluster homes. Open space is being restored to the prairie, wetland, wet prairie, and savanna communities historically found on the site. This restored landscape provides a unique living environment for the residents of Prairie Crossing. An additional 150 acres of agricultural lands are integrated into the development to protect the rural agricultural landscapes of the area.

Stormwater Management "Treatment Train System"

The open space in the Prairie Crossing project was planned to provide stormwater management for the project. The stormwater system has been designed as a treatment train with components that contribute in sequence to treat water before it leaves the site (Figure 1). Stormwater runoff from residential areas outside the village center is routed into open conveyance swales planted with native prairie and wetland vegetation, rather than storm sewers. The swales provide initial stormwater treatment, primarily infiltration and sedimentation. The prairies are the second component of the treatment train. Prairies diffuse the flows conveyed by the swales. The lessened stormwater velocities maximize the prairie's sedimentation potential. Additionally, the natural sorption sites produced by the prairie soils will hold many contaminants. The aerobic condition of the soil will also promote hydrocarbon breakdown. The prairies are expected to infiltrate a substantial portion of the annual surface runoff volume due to their porous soils resulting from very deep root systems of the prairie vegetation. Wetlands provide both stormwater detention and biological treatment prior to runoff entering the lake, which provides stormwater detention, further solids settling, and biological treatment. The components of this stormwater treatment train management system were designed to treat the stormwater runoff and reduce the stormwater runoff peaks and volumes.

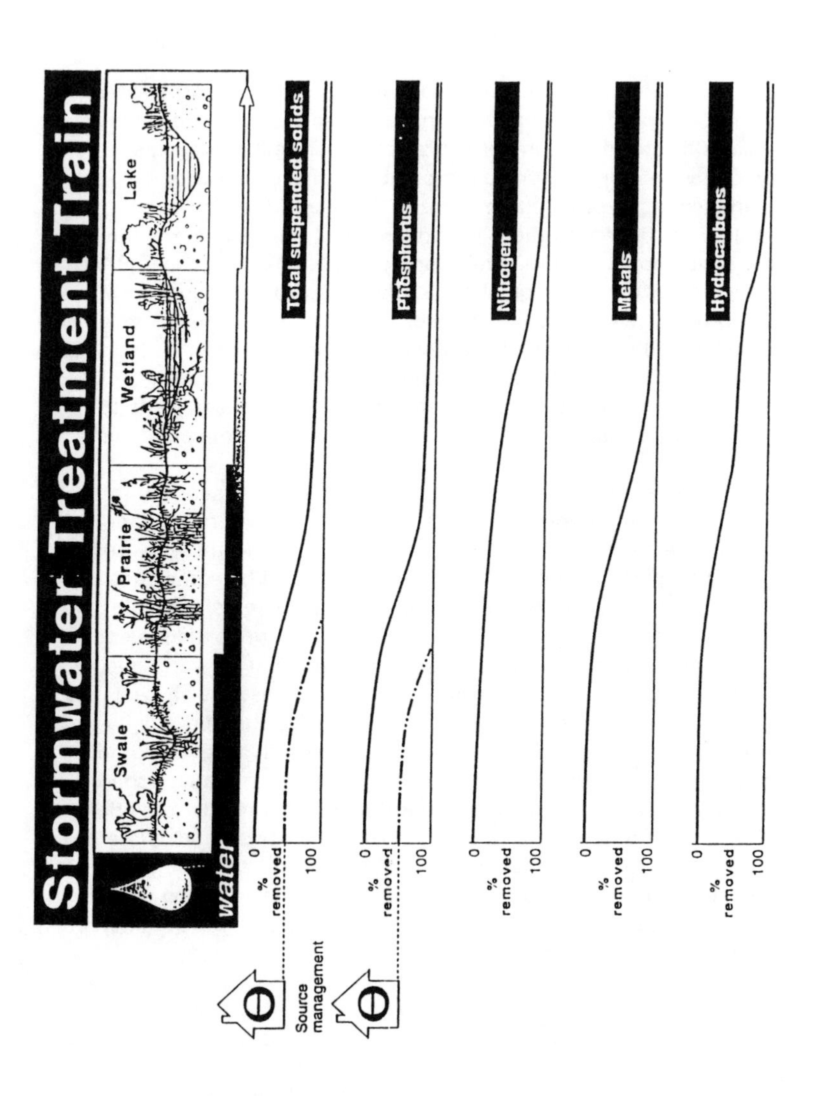

Figure 1. Functioning elements of the Stormwater Treatment Train and the anticipated general stormwater management and water quality benefits in each element.

Conclusions

The Prairie Crossing development is unique in northeastern Illinois and probably most other parts of the country. However, it utilizes cluster development and stormwater best management practices that are not unique; it combines these elements into a management system that minimizes the need for stormwater structures, enhances the living environment, and minimizes the negative impacts of urban development. Based on published BMP effectiveness information and hydrologic modeling, the Prairie Crossing development is expected to reduce surface runoff volumes by 65% and reduce solids, nutrients, and heavy metals loads by 85% to 100%. Source controls will minimize the impacts of the development even further. The result not only reduces costs to the developer, but also reduces maintenance costs for the community.

Wetlands and prairies can provide years of essentially free service, but at a great cost if mismanaged. To maintain higher quality, desirable wetlands, stormwater supplied to these systems should be higher quality, and the delivery should be somewhat predictable. What this may require is pretreatment of water in systems likely to experience extremes. Not all wetlands are equal from the perspective of biodiversity, wildlife habitat, water quality cleansing, human-use, and aesthetic perspectives. Using wetlands for water quality management can compromise the opportunity for conservation of high quality wetlands. However, this should not discount their importance and feasibility for use in water resource management. It simply means that the engineer needs to work closely with other disciplines to present accurate and adequate information for decisions makers. In conclusion, an ecological system approach, not only a focus on the importance and use of wetlands in stormwater management, is an important future direction for water resources management.

References Cited

Apfelbaum, S. I. 1997. The role of landscapes in stormwater management. Proceeding of a National Stormwater Conference on Urban runoff management, March 30-April 2, 1993. U. S. Environmental Protection Agency, Chicago, Illinois.

TRINITY RIVER, TEXAS

Moderator, Jerry Rogers

"Trinity River - Common Vision"
John Promise
North Central Texas Council of Governments

"Trinity River Basin Master Plan and Upper Trinity Water Quality Compact"
Richard M. Browning
Trinity River Authority

"Watershed Management in the Trinity River Basin"
Ken Petersen
Texas Natural Resources Conservation Commission

"Trinity River Water Supply Replacing Groundwater in Houston"
Frederick A. Pierrenot
City of Houston

"Regional Coordination for Water and Wastewater Service for Thirty Cities"*
Thomas E. Taylor
Upper Trinity Regional Water District

*Text not available at time of printing.

Trinity River Common Vision

John Promise, P.E.[1]

Abstract

The Dallas/Fort Worth metroplex is the nation's largest inland metropolitan area. Its population of 4.5 million is greater than 30 states. The Trinity River as it flows through the urban core faces great extremes, with low flows composed almost totally of treated wastewater to potential massive floods predicted to damage more than 12,000 homes and 13 million square meters (140 million square feet) of commercial property, resulting in extremely heavy flood damages.

For most of the past 150 years, the dream was that of a navigation canal to the Gulf. When the dream died in 1981, the canal was replaced by unrelated requests for federal permits to reclaim portions of the Trinity floodplain for development. Because of concern that potential cumulative impacts could not be adequately assessed through individual permit reviews, the U.S. Army Corps of Engineers, North Central Texas Council of Governments and its member local governments cooperated in a regional initiative to more comprehensively assess the problems and opportunities of the river corridor. Over the past decade has emerged a Trinity River Common Vision:

- Safe, with stabilization and reduction of flooding risks
- Clean, with fishable and swimmable waters
- Enjoyable, with recreational opportunities linked by a trail system within a world-class greenway
- Natural, with preservations and restoration of riparian and cultural resources
- Diverse, with local and regional economic, transportation and other public needs addressed

---

[1]Director of Environmental Resources, North Central Texas, Council of Governments, Arlington, Texas.

The Beginnings

"The river a little narrow deep stinking affair."

That was the first impression of A.W. Moore in 1846 upon seeing the Trinity River near present-day Dallas. For the next 150 years, it was believed that the economic future of the region depended upon navigation of the Trinity River, from Fort Worth and Dallas southward more than 483 kilometers (300 miles) to the Gulf. Thus the ultimate use of the river in the urban area was envisioned to be barge traffic with heavy industry along its banks. If some raw sewage found its way downstream towards Houston, what was wrong with that? Indeed, in 1925 the Trinity River was characterized by the State Health Department as a "mythological river of death" because Dallas led the state in deaths associated with typhoid.

In 1981, the U.S. Army Corps of Engineers officially killed the dream of navigation by determining that a federally-sponsored canal project was no longer feasible. With the Metroplex in the middle of a development boom, the Corps received numerous unrelated requests for federal Section 404 permits to reclaim portions of the Trinity flood plain for commercial and residential development.

Because of concern that potential cumulative impacts could not be adequately assessed through individual permit reviews, the Fort Worth District of the Corps and NCTCOG launched a regional initiative that has evolved through three major steps:

- Regional Environmental Impact Statement which demonstrated the need for a common approach to flood plain management
- Reconnaissance Study which identified potential flood damage reduction alternatives which could be federally cost-shared
- Feasibility Study, still underway, which is addressing flood damage reduction, recreation, water quality, environmental enhancement and coordination of other regional issues in the corridor

For the Feasibility Study, NCTCOG executed agreements with the nine cities, three counties, and two special districts along the river corridor to serve as their administrative agent. NCTCOG in turn executed a cost-share agreement with the Corps of Engineers on their behalf. Phase I of the Feasibility Study was initiated in 1990 to identify and evaluate a full range of possible projects. The six-year effort was funded at $8 million with $4 million federal, $2 million state and $2 million local cost-sharing. The work is conducted under the direction of an Executive Committee of local elected officials and senior Corps Staff, supported by a technical committee.

Emergence of Trinity River COMMON VISION

Since 1990 there have been hundreds of meetings, discussions, public forums, newsletter articles, newspaper columns, radio talk shows and TV news bites about the Trinity River. In the early 1990's they continued to pound on the problems of the Trinity -- dead bodies found along the river banks, water pollution from an industry, and so forth.

Over the last few years, the focus has changed. While there are still problems, now much of the attention is on the future of the Trinity River -- and most of it is presented as positive. Successful bond elections. River festivals. New funding for the next trails segment. Acquisition of land for preservation in the floodplain. New education programs and facilities. Editorials lauding citizen efforts. Statements by the Mayor of Dallas stressing the importance of the Trinity to the city. A new vision of the Trinity River has emerged -- a COMMON VISION --

- ***SAFE*** Trinity River, with stabilization of flood risks through the innovative Corridor Development Certificate process; restoration of levee protection in downtown Dallas and elsewhere; reduction of flooding risks through specific local/federal cost-shared projects; and greater business and homeowner participation in flood insurance
- ***CLEAN*** Trinity River, with state-of-the-art wastewater treatment facilities and coordinated stormwater programs directed at attaining fishable and swimmable waters
- ***ENJOYABLE*** Trinity River, with recreational opportunities linked by a 400-kilometer (250-mile) Trinity Trails System within a world-class greenway
- ***NATURAL*** Trinity River, with preservation and restoration of riparian and cultural resources such as the 1,400 hectare (3,500 acre) Fort Worth Nature Center, 800 hectare Lewisville Lake Environmental Learning Area, and 2,600 hectare Great Trinity Forest in south Dallas
- ***DIVERSE*** Trinity River, with local and regional economic, transportation and other public needs addressed, such as the precedent-setting Trinity Parkway Major Transportation Investment Study

The Trinity River COMMON VISION program has been recognized nationally with three prestigious awards -- the 1995 *Local Award of Excellence in Flood Hazard Management* by the Association of State Floodplain Managers; the 1996 *Achievement Award* for Major-Metro's by the National Association of Regional Councils; and in 1997 as one of the top 25 *Innovations in American Government* by the Ford Foundation.

One Example of Cooperation – the CDC Process

The Trinity River COMMON VISION program has achieved many significant milestones since its beginning. From the creation of a state-of-the-art computerized Geographic Information System to the initiation of a 400-kilometer (250-mile) world-class greenway called the Trinity Trails System, the program has demonstrated the COMMON VISION is a reality in the region. Of all these achievements, the Corridor Development Certificate (CDC) process is perhaps the most remarkable.

The Trinity River flows through nine separate cities and three counties with development permitting responsibilities, including Dallas and Fort Worth. As might be expected, each of these entities had its own separate criteria, processes and policies. Several state and federal agencies had their own requirements. In addition, the 1980's were a boom time for development with everyone looking to get their share. The detailed *Regional Environmental Impact Statement* by the U.S. Army Corps of Engineers showed that continuation of the piecemeal development actions in the corridor would make the flooding problem much worse. Therefore, stabilizing the flood risk became the top priority as the COMMON VISION program began.

It was jointly decided that common permitting criteria and procedures should be adopted. This was easier said than done. The Trinity River corridor is 620 square kilometers (240 square miles), or about 1/4 the size of the State of Rhode Island! Many months were spent debating even the simplest of topics. Sophisticated computer models were employed. Elected officials, developers, environmentalists --everyone had a voice. What emerged is indeed unprecedented. The communities improved their independent flood plain management activities by establishing a system of common development criteria and processes to be used throughout the corridor. Each of the local governments has officially amended its floodplain ordinance and adopted the CDC Manual and Process. Federal agencies such as the Corps and FEMA have bought into the process. All are making more informed and consistent flood plain decisions.

It begins with a corridor-wide computer engineering model which graphically describes in many colors the catastrophic damage that will occur when the "big flood" hits. In the computer model, more than 76,000 individual buildings have been mapped along with 0.6 meter (2-foot) interval topography, roadways, and other important details. With the improved computer technology, the Corps of Engineers has simulated a range of storm and flood scenarios, from a 2-year storm through a massive Standard Project Flood, used for planning by the USACE. The computer model and all of its data have been reviewed in excruciating detail by the engineering community - public and private. It is the foundation of the CDC process.

The CDC manual provides a consistent flood plain management tool. Under the CDC process, local governments still issue development permits pursuant to the Federal Emergency Management Agency's (FEMA) National Flood Insurance Program, but also include the additional common permitting criteria which:

- applies consistent and specific region-wide criteria, such as no rise in the 100-year flood elevation and a maximum allowable loss of valley storage in the 100-year and Standard Project Flood discharges of 0% and 5% respectively
- establishes an unprecedented Corps of Engineers review of every CDC permit request for its flood impact, well beyond the review of federal Section 404 permit requests (with a local fee system)
- allows other participating local governments along the corridor a unique thirty day period to review and comment on the development request
- provides for NCTCOG to serve as a regional record keeper through the Trinity River Information Network (TRIN) utilizing state-of-the-art GIS capabilities (and will soon be available for inspection via the world wide web)

While each individual city still makes the final development decision, it has been and continues to be well understood that a bad decision will most likely place them in court with adversely impacted communities. The CDC process reinforces 'peer pressure' by creating a common currency of engineering and permit processing. And it is working.

Success Breeds Success

Much has already been accomplished, even as the $6.85 million Phase II efforts are now focusing on implementation planning for key elements of the program, as well as exploring expansion of the COMMON VISION into the tributary watersheds. Examples from the past two years include: construction of new levees in South Dallas, and recent Dallas City Council approval of a plan that will provide relief for another 600 households; adoption of innovative "levee couplet" design for new Trinity Parkway roadway system in downtown Dallas which will also provide levee improvements for flood damage reduction; formal adoption of a 400-kilometer (250-mile) Trinity Trails System alignment by an elected officials' committee, with 65 km (40 mi) already in place and another 80 km (50 mi) in design/construction; $500,000 of funding to local governments (through NCTCOG's solid waste program) to provide regional collection of household hazardous wastes; thousands of curbside stencils warning that used oil dumping causes pollution, etc., etc.

The Trinity River is now recognized as our most important "liquid asset" as we enter the 21st century – with a new COMMON VISION.

Trinity River Basin Master Plan and Upper Trinity Water Quality Compact

Richard M. Browning[1]

The Trinity River Basin

The Trinity River watershed involves the two largest metropolitan areas in Texas. The Dallas/Fort Worth metro area, population about 4.8 million, is in the middle of the upper watershed. The Houston metro area, with 4.2 million people, is on the west side of the river's mouth at Galveston Bay on the Gulf of Mexico. There are over a dozen major water supply entities directly involved in the watershed.

Water supplies in the Trinity basin proper are predominantly from surface water impoundments, over 40 of them with a total of over seven million acre-feet of storage. Also, Dallas/Fort Worth area entities have acquired supplies in four lakes north and east of the Trinity watershed, and the lower Trinity is the largest single water source for the Houston metropolitan area. Water quality has been a problem downstream of Dallas and Fort Worth for over one hundred years, but great improvement has been achieved since 1970. Galveston Bay is the largest and most productive estuary on the Texas coast, and there is perennial interest in the impact of upstream activities on the bay.

---

[1]Senior Manager, Planning & Environment, Trinity River Authority.

Water Supply Issues and the Trinity Master Plan

A seven-year drought in the 1950s set the scene for water supply activities for many years. Even before the drought ended, all suppliers began claiming reservoir sites for future development. Conflicts occurred and in 1955 the state mandated the preparation of a master plan for the entire watershed. The *Trinity River Basin Master Plan* reconciled conflicts using long-range principles and priorities which still left suppliers free to implement their plans. The Master Plan has been revised several times with that approach, and it has been successful for forty years. Future new supplies will come predominantly from: 1) additional imports from other basins, 2) reuse of treated wastewater, and 3) conservation. Current plans for imports and reuse plus a drought in 1995-96, an array of environmental claims, and new state legislation have raised new issues and revived old ones. Handling current issues will require a new effort much like that in the late 1950s.

Water Quality Issues and the Upper Trinity Water Quality Compact

Water quality issues have been handled in a different environment: "command and control" by state and federal regulatory agencies. In order to deal with that situation, the major water agencies of the Dallas/Fort Worth area formed the *Upper Trinity Water Quality Compact*in 1975. The members are the City of Dallas, the City of Fort Worth, the North Texas Municipal Water District, and the Trinity River Authority. The compact has funded continuous water quality monitoring of 150 miles of the river for 22 years. They have conducted numerous special studies and have worked with the regulatory agencies on wasteload allocations, biomonitoring, and other issues. The Compact has minimized conflicts between local agencies and developed positive approaches to regulatory requirements. Work is underway regarding wasteload allocations in preparation for the next round of watershed permitting by the Texas Natural Resource Conservation Commission.

Watershed Management in the Trinity River Basin

Kenneth L. Petersen, Jr., Mel Vargas, and Louanne Jones[1]

Abstract

The state of Texas uses a watershed management approach for managing its water quality. The approach emphasizes collaboration with local water quality stakeholders, and will be used to develop total maximum daily loads and watershed action plans for impaired water bodies in the state, including those identified in the Trinity River Basin.

Introduction

Water quality management is a complex process involving many different players. The TNRCC, under direction from the Texas Legislature, established the Clean Rivers Program (CRP) to coordinate the efforts of the state's river authorities in managing water quality. Coordination efforts have since been hampered to some extent by the sheer size of the state and the wide differences in geographic and hydrologic conditions throughout the state. The TNRCC has implemented a watershed management approach to coordinate its efforts with those of regional and local Clean Rivers program stakeholders. The watershed management approach consolidates the state's 23 major watersheds into 5 basin groups which will be addressed in sequence for five major tasks: scoping and evaluation; data collection; assessment and targeting; strategy development; and implementation.

The TNRCC's approach calls for the development of watershed action plans that address water quality problems by evaluating all potential sources in the watershed and developing strategies for protecting or restoring water bodies. Targeting of priority watersheds is based on water bodies identified in the 1996 Clean Water Act (CWA) §303(d) List. Eleven impaired watersheds were identified in the Trinity basin, four of which have been targeted for initiation of watershed action plan development in the next two years.

---

[1] Texas Natural Resource Conservation Commission, MC150, P.O. Box 13087, Austin, Texas 78711-3087

Total Maximum Daily Loads

A critical component of watershed action plans is the development of total maximum daily loads (TMDLs). TMDLs are complex, technical assessments that estimate the maximum amount of a pollutant that a water body can receive and still meet state water quality standards. TMDLs are the scientific basis for watershed action plans. This maximum load is allocated among all the known sources of the pollutant in the watershed, such as wastewater treatment and industrial discharges, and urban and agricultural runoff. Based on this pollutant load allocation, local, regional, and state water quality managers must then implement a comprehensive watershed action plan that includes regulatory and nonregulatory actions needed to restore water quality. At every major step of the process, the public has the opportunity to participate in the development of these plans through the basin steering committees established under the Clean Rivers Program.

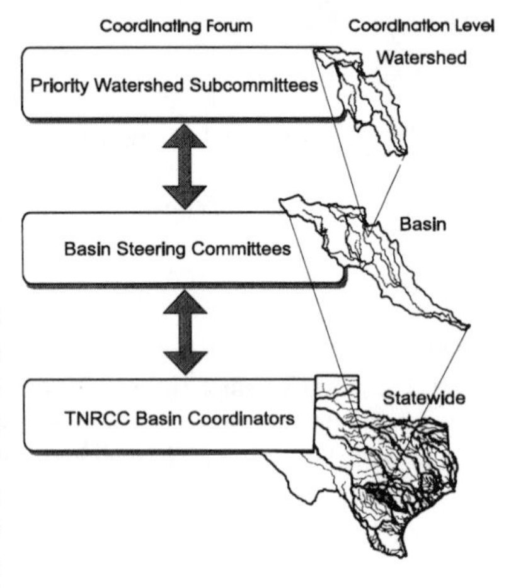

Figure 1. Forums for Stakeholder Participation

Watershed Action Plans

A watershed action plan is an implementation plan for TMDLs that identifies responsible parties and specifies actions needed to restore and protect a water body. TMDLs are the scientific basis for watershed action plans, and provide the foundation necessary to identify appropriate management objectives and strategies. Watershed action plans will document stakeholder agreements such as pollution reduction goals, pollutant load allocations, management solutions, funding options, and schedules. Plans are developed with stakeholder input through a five-phase approach as shown in Figure 2.

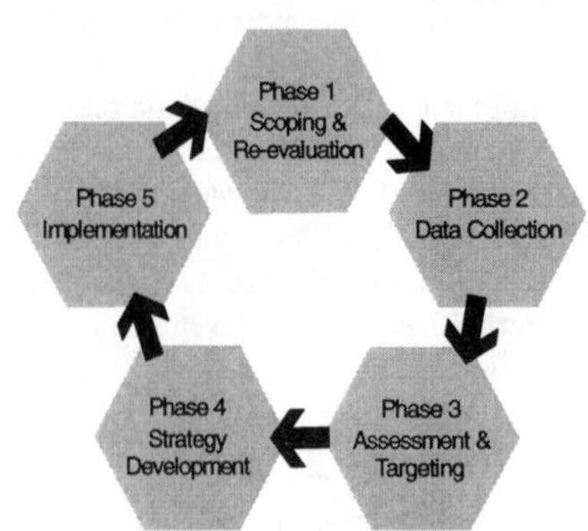

Figure 2. The Watershed Management Cycle

In Phase I, stakeholders and the TNRCC staff take a basic look at where we are and what may be needed in terms of data. In Phase II, data is gathered to determine the types of problems found in the water body and their sources. In Phase III, the data is evaluated and decisions are made about which water bodies need to be targeted for improvement. In Phase IV, management strategies and plans are developed to address the problems, and in Phase V, watershed action plans are implemented.

Trinity Basin Impairments

Eleven impaired segments were identified in the Trinity basin, four of which have been recommended for initiation of watershed action plan development in the next two years. The segments whose watersheds are targeted for TMDL development are:

#806, West Fork Trinity River below Lake Worth,
#829, Clear Fork Trinity River below Benbrook Lake,
#831, Clear Fork Trinity River below Lake Weatherford, and
#841, Lower West Fork Trinity River.

Impairments in these segments include aquatic life closures due to chlordane and other toxics in fish tissue, nonsupport of contact recreation due to elevated levels of fecal coliform bacteria, and nonsupport of aquatic life use due to elevated metals and depressed dissolved oxygen levels. A TMDL must be developed for each pollutant causing nonsupport in the affected segments.

The TMDL Development Process

There are six major steps in the development of the TMDL:

- Involve the public at the watershed level,
- Select candidates for TMDLs from List of Impaired Water Bodies,
- Develop loading limits on certain pollutants,
- Produce TMDL document for EPA approval,
- Produce and implement a watershed action plan to restore the water body, and
- Monitor water quality to gauge attainment of water quality standards.

The development of TMDLs in the Dallas/Fort Worth metroplex will require significant collaboration between TNRCC and local stakeholders. The primary forum for stakeholder input will be the Clean Rivers Program basin steering committees. This public involvement will be encouraged at each major step in the process.

Once the scoping phase has been completed (initial public comment and selection of candidates), data must be gathered to more accurately determine the sources of each specific pollutant impacting the water bodies, along with more specialized information about the impairments. Using this data, water quality professionals will use modeling and assessment tools to develop total maximum daily loads for pollutants of concern (assessment and targeting phase). The scientific rigor and equitable allocation of pollutant loads will be largely dependent on the availability of in-stream water quality data collected by numerous agencies in the Dallas/Fort Worth metroplex and the consistent participation of stakeholders. These loads will then be discussed in a report that details the pollutants of concern, the sources of pollution, and the methods for developing the load allocation. On approval of the TMDL by the Environmental Protection Agency (EPA), a watershed action plan will be developed (strategy development phase) and implemented to achieve the pollutant load reductions required under the TMDL. Watershed action plans will address both point and nonpoint sources of pollution using an appropriate combination of regulatory and voluntary approaches.

While point and nonpoint source pollution controls are currently in place throughout the watersheds, additional controls and management measures are expected to be implemented on completion of the TMDL process in fiscal year 2001.

A Joint Effort

The effort required to produce even one TMDL is enormous, and will require the cooperation and hard work of people at all levels of the government and private sectors. Local understanding and support for pollution controls will be necessary to reduce the pollutants of concern and to restore these impaired water bodies.

TRINITY RIVER WATER SUPPLY REPLACING GROUNDWATER IN HOUSTON, TEXAS

Frederick A. Perrenot [1], P.E.

Abstract

Houston is in transition from primary reliance on groundwater supply to a surface water system, a transition which began in the mid-1970's and will not be completed until 2015. This change is necessary to control coastal land subsidence in Galveston and Harris Counties, where valuable property has been lost. This transition includes a major program of new treatment facilities and interconnection transmission mains plus additional surface water supplies. Groundwater will remain as a usable resource, but pumpage will be limited to less than 200 million gallons a day. Houston operates over 110 separate groundwater pump stations which originally had minimal interaction with each other. Legally, implementation of the surface water conversion program is mandated by the Harris-Galveston Coastal Subsidence District, a political body created by the Texas Legislature. Politically, there are many issues to settle regarding water rates to other cities and to gain required cooperation among the various governmental agencies. Houston diverts Trinity River water via a pumping station and canal to the Lynchburg Reservoir and Pumping Station. The quality of the Trinity river source is important to Houston and the lower trinity river water users.

From Surface to Groundwater

The current program of conversion to a water supply system that mainly relies on surface water is reversing the trend of Houston reliance on ground water that began in the late 1880's. At that time water was supplied to Houston by the privately owned system of the Houston Water Works Company. Until 1887 the major source of water was Buffalo Bayou which at that time yielded apparently, relatively clean water.

In 1906 the City purchased the system and began a program of improvements to the system that included total conversion to well water, metering of water use, and extension of water mains "to keep pace with the City's unusual development". When the City purchased the water supply system, ground water pumpage was increased to about 11 mgd.

The plentiful supply of high quality ground water led to large scale withdrawals for public supply and rice irrigation in the Houston-Galveston region beginning the late 1800's. After the Houston Ship Channel was opened to ocean-going vessels in 1915, oil refineries requiring large amounts of water were constructed in eastern Harris County, contributing to rapid growth in ground water consumption. By 1940 ground water use had risen to 154 mgd. The use of ground water increased rapidly in the early 1940's as a result of the water needs of

---

[1] General Manager, Public Utilities Group, Department of Public Works and Engineering, City of Houston, 1801 Main Street, Houston, Texas 77002

major war industries locating in the region. It continued to increase into the early 1950' s as a result of the region's growing population, rice farming and industry. This growth in ground water use continued until 1954 when Lake Houston, the West Canal, and the first phase of the East Water Purification Plant were placed in operation.

Symptoms of Groundwater Problems Develop

Due to the almost total reliance on ground water in the Houston-Galveston region, the symptoms of overuse began to be observed in the 1940's and 1950's. Groundwater quality began to deteriorate. Dramatic drops in the water levels occurred in both major aquifers in the region (the Chicot and Evangline). Land-surface subsidence were beginning to be recognized as a major problem in the low-lying area near Galveston Bay and the Gulf of Mexico.

Ground water pumpage for the region increased from about 154 mgd in 1940 to about 388 mgd in 1954 when major supplies of surface water became available. Ground water usage remained relatively stable until 1962 (398 mgd) and then again increased dramatically to 507 mgd by 1969.

The declines in artesian pressure have resulted in land-surface subsidence of more than one (1) foot in an area of about 3,640 square miles. From 1906 to 1987 subsidence of about nine (9) feet had occurred along the Houston Ship Channel with maximum subsidence of about ten (10) feet in the Pasadena Area in the eastern part of the region.

The Beginning for A Solution

Water planners for the City of Houston determined that further surface water sources needed to be developed. This included Lake Livingston completed in 1969 in partnership with the Trinity River Authority. In addition, Lake Conroe was completed in 1973 in partnership with the San Jacinto River Authority. Plans to build a saltwater barrier on the lower Trinity River were also developed in order to maximize the safe yield of the Trinity River system.

To utilize the new surface water source being developed at Lake Livingston, the Coastal Water Authority (formerly the Coastal Industrial Water Authority) was created by a special act of the Legislative in 1967 to develop a conveyance system from the Trinity River to deliver raw surface water to the Houston Area. As a part of the conveyance system for industry and irrigation, the Southern Canal System was purchased in 1969.

By 1976 CWA had completed the Trinity River Pumping Station, the Main Canal, Lynchburg Reservoir, the Lynchburg Pumping Station, the 108-inch Pipeline Ship Channel Crossings, and the Pipeline System to Ship Channel Industries. Surface water was first provided to these industries that year.

Subsidence District Created

The Harris-Galveston Coastal Subsidence District (HGCSD) was created in 1975 by the 64th Legislature to regulate the withdrawal of groundwater within Harris and Galveston Counties. The District was created "for the purpose of ending subsidence which contributes to or precipitates flooding, inundation, or overflow of any area within the district including, without limitation, rising waters resulting from storms or hurricanes".

In 1976 HGCSD adopted an interim plan designed to have an impact on the subsidence problem in the area most vulnerable to the damaging effects of subsidence. Much progress was made in controlling subsidence in this area of the Southeastern part of Harris County and all of Galveston County, between 1976 and 1984 largely due to the provision of raw surface water to the Galveston area from the East Water Purification Plant. Water is also provided to the Galveston Area by the Galveston County Water Authority.

1985 to 1992 Subsidence Regulation

By 1985 the rate of subsidence in southeast Harris and Galveston County had dramatically declined as the direct result of reducing groundwater withdrawals. In addition, water level increases in the aquifers in amounts up to 150 feet had occurred. However, surface water was not available in the western and northern parts of Harris County. This lack of supply, coupled with extensive growth, had allowed subsidence in those areas to continue at a rather rapid pace.

By 1985 the Subsidence Plan adopted by the HGCSD Board of Directors divided the region into eight regulatory areas, and prescribed the date for each area to convert to surface water use and the percentage of use to be made up of surface water.

In 1989 the Fort Bend Subsidence District was created to regulate groundwater withdrawal in the County which is southwest of Harris County and includes a part of the City. From data available at that time, it appeared that southwest Harris County and northeast Fort Bend County were the primary areas where subsidence would occur based on projected groundwater demand.

More recent data indicates that the growth of water demand in southwest Houston and northeast Fort Bend County is substantially less than previously predicted and the time frame and areas for conversion in Harris County has been revised. In 1992 the HGCSD Plan was revised, which resulted in the number of regulatory areas being reduced to seven and the focus for earlier conversion was shifted from southwest to northern Harris County.

City of Houston Programs

In 1985 the City retained an engineering consulting firm to manage the program for the development of the surface water transmission system. This program was designed to assist the City in meeting the surface water conversion dates established by the HGCSD. From the start of this program through 1997 the City has built or started construction on about 73.5 miles surface water transmission lines ranging from 36-inch to 96-inch diameter, and an additional 32 miles of water lines ranging from 12-inch to 30-inch diameter.

The City annually prepares and updates a Five Year Capital Improvement Plan for the water system. The City's consultant has also started developing the conceptual plan for surface water conversion for those portions of all subsidence areas that are located within the City. Due to the configuration of the City limit lines in northern and western Harris County the plan, of necessity, is developing a concept for providing surface water in the areas outside the City limits.

In order to supply the required amounts of surface water within Regulatory Areas One through Three, the City has also carried out a program of expansion and new construction of its surface water treatment plants. In 1987 Phase III of the East Water Purification Plant was completed giving it a treatment capacity of about 300 mgd.

In 1990, Phase I of a new Southeast Water Purification Plant with an average day capacity of 50 mgd and a maximum of 80 mgd was completed to provide regional service to serval entities in order to meet subsidence criteria for that area. The financing of the construction of this plant was a departure from previous City plant construction. The City solicited participants in a cost sharing plan for the construction, operation and maintenance of this plant. There are nine (9) participants and the City, with each participant having a different share of the plant capacity. The contracts also provide for the participants to be included for additional capacity that may be constructed. The City currently has a 22% share of the plant capacity.

Because of the multiple capacity ownership, the participants contracted with an outside firm for the maintenance and operation of the Southeast Plant. The experience gained in developing this plant with outside participation and contract operation will be of great value in the future construction of new water purification plants.

In 1974 the City of Houston determined that another water purification plant located near Lake Houston might be necessary to provide surface water for the northern portions of the City and Harris County. This led to the planning and site acquisition for the Northeast Water Purification Plant.

In conjunction with the plans for the Northeast Water Purification Plant, it became apparent that ultimately additional water, beyond the 130 mgd available in Lake Houston and the City's 60 mgd share in Lake Conroe, would be needed in the San Jacinto River System to supply the needs of the Northeast and East Water Purification Plants. Planning was carried out for a project to divert water from Lake Livingston to Lake Houston taking advantage of the location of Luce Bayou, one of the tributaries to Lake Houston. Again, plans were developed and a Texas Natural Resource Conservation Commission (TNRCC), at that time the Texas Water Commission, permit was obtained. The addition of added capacity to the system for delivering water from the Trinity River to the East Water Plant by completion of the Northwest Lateral in 1991 will allow the East Water Plant to operate, using Trinity River water only, for a significant period of time.

Costs of Conversion to Date

The City of Houston has a forty year history in the conversion from groundwater to surface water use and during that time has assumed the role of the regional supplier. While the achievements have been many, the price has not been cheap. To date the City and its rate payers have invested, in actual unadjusted construction costs, over $1.8 billion of principal and interest.

This has had an immense impact on the City water rate payers. In 1983 the monthly water bill for a customer using 7,000 gallons per month, about the average use, was $11.34. In March of 1993 the same customer was paying $20.47 per month, an 80.% increase in 10 years. Unfortunately, the end is not yet in sight.

Water Planning

In 1985 a consultant was selected to develop a water master plan that looked forward to the year 2030. The plan concluded that without the Trinity River saltwater barrier a new source of water would be needed in about 2010 and with the barrier about 2030. The existing available source for additional large amounts of water was identified as the Toledo Bend Reservoir on the Sabine River.

In May of 1992, the Mayor of Houston convened a Regional Water Futures Assembly that was held in Houston and invited the Mayors of San Antonio and Corpus Christi, the Chairman of the Texas Water Development Board and other interested parties to participate in developing a plan for the transfer and use of surface water in the southeast and south central areas of Texas. This meeting set the stage to begin the Trans Texas Water Program. This program was established to conduct a feasibility study for the transfer and exchange of water and water rights from the Sabine River on the eastern boundary of Texas as far west as Corpus Christi and San Antonio. In February 1993 the City of Houston entered into a joint sponsorship agreement with the Sabine River Authority and the San Jacinto River Authority for the Southeast Study Area of the Trans Texas Water Project. This study determines the feasibility of the transfer of Sabine River water to the Houston area.

Integrated Resource Planning and Management

We visualize meeting the needs of the future to start by maximizing the use of existing available resources. This requires that we develop a plan that takes advantage of the great groundwater resource available to us. Conversion for subsidence does not require that we turn completely to surface water. The Houston Water Master Plan determined that about 350 mgd of groundwater would still be available for use in the year 2030 when total compliance with the subsidence requirements had been achieved. The Trinity River Salt Water Barrier is currently under construction. When completed it would make available about 260 mgd of raw water by eliminating fresh water releases to prevent saltwater intrusion and by perfecting the water rights assigned to the project. Those water rights are based on a system yield from the combined Lake Livingston Salt Water Barrier operation.

As we progress in the development of presently planned projects, the system operation of the existing available reservoirs will allow us to make maximum use of available water particularly in periods of prolonged drought. Currently, we need to develop a plan for the system operation of Lake Conroe and Lake Houston. With completion of the Saltwater Barrier, a plan for system operation of Lake Livingston and the Barrier will be developed. Finally when the Luce Bayou Diversion is constructed, the three reservoirs and the barrier can be operated as a system.

A plan for the conjunctive use of groundwater and surface water needs to be developed. The timing for operation of wells and treatment plants can be managed to maximize the use of available water sources. The location of wells and groundwater plants to be retained after subsidence conversion needs to be determined for maximum benefit.

Conservation was considered as a source of water by the Houston Water Master Plan. The reduction in consumption in effect makes water available for future growth. While conservation is an essential part of the planning, unrealistic reliance on the reduction of water use must be avoided. Water Reuse is another area that will become practical as the cost of providing water increases. We are in fact already "reusing" treatment plant effluent that is discharged into the San Jacinto and Trinity River systems above the reservoirs that serve our water plants. Water utilities should have a great interest in selling the same water twice.

There will need to be significant increases in the capacity of the Southeast and proposed Northeast Water Plants. The output of the existing East Water Plant will need to be maximized and/or expanded. other locations for water purification plants may be identified and utilized. The Houston Water Master Plan included a Southwest Water Plant using Brazos River water. Depending on the location that is selected for delivering water from the Sabine River in the Trans Texas Water Project other water plant locations may be identified and utilized for maximum benefit.

The transmission system for delivery and storage of treated surface water will be vastly expanded to serve the existing developed areas. Further growth in the areas not served by the system will by necessity need to make provision for connection to the system. The Houston Water Master Plan identified a Regional Surface Water Service Area that grew over time to include those areas where groundwater use had reached the maximum allowable. This is a sound concept and though the timing for development of the system has and will continue to change, that growth in system needs consideration in the current planning.

To date our accomplishments have been great, but many challenges remain. As the regional. water supplier the past conversions to surface water use has been successful largely because of the comparatively limited number of entities that were participants in the program.

To date no single entity has been identified that can finance a project of this magnitude with such a large number of participants. The method of equitably distributing the cost among the users will have to be developed. This must be done in a manner that does not give an unfair advantage to any of the participants.

Study should also be done to sort out all of the alternatives and select the most practical solution. While the technical design of the system must progress to the point where the magnitude of the cost can be determined, it appears that the real thrust of the study should be directed at the legal, institutional, political and financing issues. These are the issues that need to be resolved before the implementation of a technical plan can be carried out.

The City sees itself as having a role in each of the areas of regulator, supplier and distributor/user. The City is a regulator in the review of plans and the creation of utility districts. The new water conservation plan rules of TNRCC may require that we review individual entities plans for compliance with the rules.

I leave you with the thought that over the years there have been visionaries that have given the Houston area its current system of surface water sources and defined the limits of our valuable groundwater resource. I only hope that together the Houston area may develop and implement the planning for the future in a manner that those persons responsible for the system in 2030 will be saying "the water planners for the Houston area water system in the 90's, with almost unbelievable foresight ....

CONTROL OF MUNICIPAL AND INDUSTRIAL RUNOFF

Moderator, Scott Tucker

"Making Watershed Management Work for the National Pollutant Discharge Elimination System"
Michael B. Cook
U.S. Environmental Protection Agency

"Municipal Stormwater NPDES Permit Issues"
Lori L. Sundstrom
City of Phoenix

"Stormwater Pollution Prevention – Compliance Assessment of Industrial Facilities in California"
L. Donald Duke and Kathleen Shaver
UCLA School of Public Health, Los Angeles

Making Watershed Management Work for the National Pollutant Discharge Elimination System[1]

Michael B. Cook, Gregory W. Currey, and Ben Lesser[2]

Abstract

The Clean Water Act and its implementing regulations attempt to control urban wastewater flows (continuous flows from publicly owned treatment works and intermittent wet weather flows such as storm water from municipal separate storm sewer systems, storm water from industrial sources, combined sewer overflows or CSOs, and sanitary sewer overflows or SSOs) by regulating them as separate "point sources" of pollution under the National Pollutant Discharge Elimination System (NPDES). This approach leads to the iterative imposition of multiple overlays of monitoring and control requirements, each addressing a different pollution component. In addition, some significant pollution sources may be overlooked.

The Agency is considering a framework for "watershed permitting" as an alternative to separate permitting of individual pollutant discharge sources. In urban areas, such a permitting system would include a coordinated approach to addressing many sources in one permit or a few permits, including both urban wet weather and continuous sources of pollution. Where this system of permitting is adopted, it will help to highlight the most critical pollution control needs of a watershed and community and allow those needs to be addressed in the most effective and least costly manner consistent with the goals of the Clean Water Act.

Introduction

One of the largest and most successful regulatory programs of the U.S. Environmental Protection Agency (EPA) is the National

---

[1]The views presented in this paper are solely those of the authors and do not necessarily reflect the views of the U.S. Environmental Protection Agency (EPA) or any other Federal agency.

[2]Michael B. Cook is Director of the Office of Wastewater Management, EPA. Gregory W. Currey is a Civil Engineer with the Office of Wastewater Management, EPA. Ben Lesser is Wet Weather Teams Coordinator in the Office of Wastewater Management, EPA.

Pollutant Discharge Elimination System (NPDES). Congress created NPDES through the Clean Water Act[3], and the program is implemented by EPA's Office of Wastewater Management (OWM). Since 1972 this program has grown to include regulation of nearly 70,000 "point sources" of wastewater discharge from industrial processes and municipal sewage treatment plants to surface waters of the United States. NPDES requires dischargers to obtain permits setting appropriate technology-based and water quality-based pollution limits in the wastewater, or effluent, they discharge. EPA Regions or authorized states develop these limits and issue permits, which are enforceable under Federal and state laws. OWM develops NPDES program regulations, policy, and guidance, provides technical support for the program, and funds projects to demonstrate new, more effective and efficient ways to comply with the law.

After more than two decades of NPDES implementation, EPA studies and other reports show that the most visible and immediate threats from regulated point sources have been brought under control. For example, rivers no longer catch fire, as debris in Ohio's Cuyahoga River once did, fish have returned to many rivers and lakes, and the public generally supports the costs of addressing major threats to U.S. waterways through the NPDES program.

<u>New Challenges: Wet Weather and Watershed Management</u>

Successfully controlling many of the most immediate pollution threats to the nation's waters has allowed a new challenge to clearly emerge. According to state reports, 36 percent of surveyed rivers and streams, 37 percent of surveyed lakes, and 37 percent of the estuaries surveyed by coastal states are partially or fully "impaired," meaning that they do not meet State standards for their use (Office of Water, 1995). Pollutants such as silt, nutrients, bacteria, metals, pesticides, and organic chemicals are found in impaired waters. "Wet weather flows" of storm water, sewage overflows, and runoff from lawns and farm fields, mining, and forestry contribute significantly to the loading of these pollutants in receiving waters. Wet weather flows result in beaches closed to swimming, shell fishing areas closed, reduced spawning of salmon, fish kills, and increased costs to treat water for drinking or other uses.

In the 1980s, it became clear that the NPDES program needed to address this new challenge. Congress, through the 1987 Water Quality Act (which reauthorized the Clean Water Act), emphasized that the NPDES program should be broadened to include more diffuse sources. In the early 1990s, EPA developed regulations to control stormwater from large municipalities and certain industrial activities. EPA's *National Water Quality Inventory, 1994 Report to Congress* confirmed the need for wet weather controls, showing roughly 46 percent of water quality impairment to be attributable to storm water discharges (Office

---

[3]*Federal Water Pollution Control Act Amendments of 1972 (P.L. 92-500)*, more commonly called the Clean Water Act.

of Water, 1995). Thus, in addition to point source discharges from industrial processes and municipal treatment plants, the NPDES program now covers tens of thousands of other discharges from stormwater and other diffuse sources such as concentrated animal feeding operations and combined sewer overflows (CSOs).

Faced with an overwhelming number of pollutant sources subject to regulation under the Clean Water Act, in 1993 EPA embarked on a major re-direction for the entire NPDES program. The Agency recognized the importance of addressing all pollution sources within a hydrologically-defined drainage basin, or watershed, in an integrated manner instead of viewing individual pollutant sources in isolation. Also, the agency recognized states, American Indian tribes and other NPDES program operators as essential partners in supporting and achieving many of the Clean Water Act's water quality objectives. Working from these principles, the 1994 NPDES Watershed Strategy established two major goals: 1) to integrate NPDES program functions for the national water program into a watershed management framework and 2) to support development of state-wide watershed management approaches (Office of Water, 1994).

Since developing the NPDES Watershed Strategy, EPA and the states have made a considerable effort to meet these two goals. EPA has developed and is continuing to develop policies and guidance to ensure that core program functions (e.g., effluent limit development, timing of permit issuance) consider whole watersheds and interactions among pollutant sources within watersheds. EPA also has provided technical and financial assistance to a number of states that have developed watershed management frameworks for operation of their water resource programs. Finally, EPA has developed a framework that provides for equal consideration of all eligible water quality projects for State Revolving Fund assistance, including wastewater, nonpoint source, and estuary enhancement projects, using existing water quality and watershed information.

EPA now has the challenge of continuing to advance the NPDES program using watershed management principles to address industrial and municipal wastewater discharges, including wet weather flows, while strengthening and maintaining the core of the existing NPDES program: technically sound, enforceable permits that attain technology-based performance requirements and meet state-defined water quality standards for the use of a particular water body. The Agency is developing a permitting framework that looks toward innovative and flexible solutions to watershed management issues that will improve environmental quality, but may be beyond current common practice within the NPDES program. The following, sections describe two important areas of work that look toward meeting this challenge.

Advisory Committee Guidance and the Watershed Alternative

On May 1, 1995, EPA announced the establishment of the Urban Wet Weather Flows Federal Advisory Committee under the Federal Advisory Committee Act. Members include 32 stakeholders from 16 states and the District of Columbia representing municipalities, industries, environmental groups, public interest groups, state government, tribes, professional associations, small businesses, health officials, conservation interests, and EPA. The Agency asked the Committee to help develop recommendations for controlling the environmental and human health effects of urban wet weather flows, with a minimum of regulatory burden.

The Committee soon agreed that a watershed-based pollution management system provides important benefits including:

- greater opportunities to improve water quality
- a more equitable allocation of water quality management responsibilities
- an improved basis for decision-making
- enhanced efficiency and lower costs
- improved coordination
- an emphasis on the local role in identifying problems and solutions
- increased opportunities for market-based approaches
- improved relations among stakeholders
- enhanced public involvement

In the 438-square mile Rouge River, Michigan watershed, 15 communities are working cooperatively to control pollution from their combined sewer overflows (CSOs) as part of the Rouge River National Wet Weather Demonstration Project. Preliminary data indicate that CSO impacts can be controlled effectively and at much less cost by addressing overflow points in the context of an overall watershed management plan implemented to address all sources of impairment in the river system. Such a watershed approach facilitates individual control facilities tailored to the needs of the watershed, as an alternative to using a single design for all CSO control facilities in a geographic area. CSO control project costs, originally estimated at approximately $2.5 billion for the Rouge River, may be reduced to $1. 5 billion. The potential cost savings of about $1 billion can be used to address other Rouge River watershed pollution problems (Personal communication Kelly Cave. 1997).

The Committee developed the *Watershed Alternative for the Achievement of Water Quality Objectives,* which highlights the benefits of pursuing a watershed-based decision making process, components that should result from the process, and actions that EPA and state regulatory authorities can take to encourage NPDES permittee participation in the process, including incentives and a mechanism for addressing non-NPDES contributors to watershed pollution (Office of Water, 1997). The Committee also is preparing recommendations on how to develop watershed-based assessment strategies.

<u>NPDES Watershed Permitting</u>

Although the *Watershed Alternative* represents a framework for watershed planning, some communities are seeking innovative ways to take full advantage of a watershed approach to implement water quality management actions. As a result, EPA is exploring models that could be used for implementing a watershed-based permitting program. Watershed permitting models that EPA is examining include: common or related permit conditions in multiple NPDES permits within a watershed; general permits used in a watershed to address a single category of point sources with similar discharges; general permits used to address multiple types of discharges by focusing on a parameter(s) of concern within a watershed; and, a single watershed permit held by a multi-jurisdictional authority or "umbrella" organization. Each model provides a regulatory and administrative mechanism that could be used to address multiple point sources within a watershed or other geographic area. Some also could incorporate management measures affecting important, but unregulated pollution sources and other factors that put "stress" on a watershed. These sources could be addressed either through permit requirements or through mechanisms for pollutant trading between regulated and unregulated sources.

The goal of watershed permitting is to provide a tool that will allow communities and permitting authorities to determine the most effective and efficient means of achieving water quality standards and locally defined watershed objectives, and to achieve watershed goals that they could not meet otherwise. To demonstrate the benefits of a watershed permitting framework, EPA is conducting case studies in several interested communities. These case studies will focus on how a watershed permit would be developed and the expected benefits to the watershed and the community at large. The case studies will help the Agency re-examine its current permitting framework and will lead to watershed permitting pilot projects that serve as models for other communities and permitting agencies.

The Louisville and Jefferson County, Kentucky's Metropolitan Sewer District (MSD) manages municipal wastewater and storm water and oversees the industrial pretreatment program throughout Jefferson County. Seven watersheds lie fully within the District's jurisdiction, and four more lie partially within it. Within those seven watersheds, MSD holds 43 point source discharge permits issued by Kentucky). Despite infrequent permit violations, no Jefferson County streams support recreational use and only a small percentage of local streams support warm water aquatic life uses. Watershed permitting could allow MSD to consolidate its existing 43 point source permits into 7 watershed permits. This approach could streamline the permitting process; provide a framework for watershed-based monitoring, data collection and effluent controls; allow MSD to set priorities for investments based on environmental outcomes, and create a forum for concurrently addressing regulated and unregulated sources of pollution in Jefferson County watersheds (Apogee Research, Inc., 1997).

In addition to the case studies, EPA is developing a

watershed permitting program analysis that will describe a range of possible watershed permitting models and outline changes or amendments to existing national regulations, policy, and guidance that would facilitate their implementation. Based upon the watershed permitting program analysis and lessons learned from the community case studies and pilot projects, EPA will develop and propose a national watershed permitting framework. The framework may include regulatory amendments, policies, and guidance on developing and implementing watershed permits. EPA will seek public comment on many of these components.

Implementing the Watershed Framework

The *Watershed Alternative* decision making process, the guidance on developing watershed-based assessment strategies, and the watershed permitting mechanism are key products being developed for the NPDES program's national watershed framework. Together, they promise numerous benefits to permitting authorities, permit holders, and watershed communities alike, including elimination of multiple overlays of monitoring and control requirements, better measurement of water quality and effectiveness of pollution controls, and much greater cost-effectiveness by coordinating controls on multiple watershed pollution sources. We believe the intelligent use of these tools will improve the nation's ability to protect the health of its vital waters, and will enhance the quality of life of its citizens.

References

Apogee Research, Inc., "Preliminary Findings: Louisville and Jefferson County, Kentucky Metropolitan Sewer District and Watershed Permitting," Memorandum from Apogee Research, Inc. to U.S. Environmental Protection Agency, September 30, 1997.
Office of Water, *NPDES Watershed Strategy, U.S.* Environmental Protection Agency, Washington, DC, March 21, 1994.
Office of Water, *National Water Quality Inventory 1994 report to Congress, U.S.* Environmental Protection Agency, Washington, DC, December, 1995.
Office of Water, *Watershed Alternative for the Achievement of Water Quality Objectives, U.S.* Environmental Protection Agency, Washington, DC, Draft, November, 1997.
Personal communication with Kelly Cave, Director, Watershed Management Division, Wayne County Department of Environment, December 18, 1997.

Municipal Storm Water NPDES Permit Issues

Lori L. Sundstrom[1]
Craig J. Reece[2]

Abstract

The City of Phoenix, Arizona (Phoenix) received a National Pollutant Discharge Elimination System Permit (Permit) from the U.S. Environmental Protection Agency, Region IX, on February 14, 1997. The permit requires implementation of best management practices, storm water quality monitoring, and annual reporting to EPA. The permit does not contain numeric discharge limits, consistent with current EPA policy on first generation municipal storm water permits. This article discusses the major practical and legal issues considered during the permit's negotiation.

Background

Congress created the National Pollutant Discharge Elimination System (NPDES) Program to restore, protect and maintain the biological, chemical and physical integrity of the nation's waters. The central mechanism used to achieve these goals is the NPDES permit which establishes enforceable requirements that articulates the amount of pollutants a point source may discharge to a water of the United States. NPDES permit limits or impose other conditions on what then becomes a lawful discharge. In the 1987 amendments to the Clean Water Act (CWA), Congress directed the U.S. Environmental Protection Agency (EPA) to expand its NPDES program to include a storm water component.

The City of Phoenix applied for an NPDES permit for its municipal separate storm sewer system (MS4) in 1992. Its storm drain system includes approximately 1,900 miles of pipe, 600 miles of open channels, and 153 major outfalls.[1] Upstream

---

[1] Environmental Affairs Supervisor, City of Phoenix, 200 West Washington Street, Phoenix, AZ 85003-1611

[2] Assistant City Attorney, City of Phoenix, 200 West Washington Street, Phoenix, AZ 85003-1611

dams long ago impounded the metropolitan area's rivers. Now natural channels typically carry flow only during periods of runoff and dam releases. The Phoenix metropolitan area receives a mean annual precipitation of approximately seven inches per year.

On February 14, 1997, Phoenix received an NPDES permit from EPA's Region IX. This permit requires Phoenix to implement a Storm Water Management Plan (SWMP), which Phoenix developed and EPA approved, and which contains thirty-six best management practices (BMPs) designed to improve urban storm water quality. Phoenix is also required to monitor storm water quality, assess the effectiveness of the BMPs, and report its activities annually to the EPA. The following discussion briefly describes the major issues considered in negotiating Phoenix's MS4 storm water permit.

Applicability of Water Quality Standards

Implementing the current NPDES program framework in the context of municipal storm water discharges raises several important practical and legal issues. Desert urban storm water does not lend itself well to characterization and measurement, largely due to its dynamic nature: rainfall is infrequent and episodic, and its composition varies within and between storm events. The traditional reliance on water quality criteria is therefore problematic for desert urban storm water. Fr example, an exceedance of a numeric water quality standard may occur for only a short time during or immediately after a storm, and it does not necessarily mean that the water quality of the receiving stream has been impaired. It is difficult to measure instream impacts in ephemeral streams. Consequently, it is difficult to relate storm water quality data to the effectiveness of municipal storm water management plans.[2]

Other issues arise from the fact that the quality of the MS4's discharge is the aggregate of discharges from many sources in the urban watershed. It is practically impossible to distinguish the effect of one source's discharge from another, while at the same time it is necessary to apportion responsibility among all of the contributors to the discharge. EPA lacks authority to impose numeric effluent limitations on MS4 outfalls because doing so would shift legal liability for discharges by third parties to the municipal permittee. At the heart of the matter is the fact that the EPA is treating municipalities differently from the other point sources. For all other point source categories, the EPA regulates individual facilities. In the Phase I MS4 rules, EPA recognized that individual facilities that used the MS4 were individually responsible for their storm water discharges. In its recently proposed Phase II MS4 rules, EPA is retreating from this position and is seeking to treat the MS4 as a facility and therefore the source of the discharge. This is inappropriate for reasons explained below.

The CWA specifies that permits for MS4s shall require controls to reduce the discharge of pollutants "to the maximum extent practicable."[3] The EPA has also concluded that MS4s must comply with water quality standards, and the point of compliance is the outfall. The basis for this logic is expressed in a January 9, 1991 memorandum from the EPA Office of General Counsel. The General Counsel's opinion did not, however, imply that the MS4 owner is responsible for water quality violations caused by private facilities.

EPA also recognizes that, "the CWA does not say that effluent limitations need be numeric. As a result, EPA and States have flexibility in terms of how to express effluent limitations."[4] An August 26, 1996 EPA policy, "Interim Permitting Approach for Water Quality-Based Effluent Limitations in Storm Water Permits," provides guidance to EPA Regions and delegated states that allows first generation permits to deem implementation of required BMPs as compliance with water quality standards. Phoenix's permit is consistent with this approach, and does not contain numeric limits.

Responsibility for the Actions of Third Party Dischargers

The fundamental issue is whether the CWA and the U.S. Constitution allow the EPA to require conditions in NPDES permits for MS4s that make the MS4 owner responsible for storm water runoff from non-municipal facilities. The MS4 is not legally responsible for discharges by third parties who are independently subject to other NPDES permits and other CWA legal requirements.

There are four reasons why EPA legally cannot and should not impose permit limits that make the MS4 operator responsible for the quality of discharges from other sources. First, the CWA apportions regulatory authority between the EPA and states with approved NPDES programs and does not provide for municipalities to regulate storm water discharges under authority of the CWA. Second, the CWA does not allow EPA to impose permit conditions on one permittee in order to control discharges by unrelated third parties. Third, the U.S. Constitution prohibits the Federal Government from compelling local governments to operate Federal regulatory programs.[5] Finally, making municipal system operators responsible for discharges by third parties vitiates the municipal programs required by section 402(p) of the CWA.

A central tenant of this argument is that the owner of the MS4 is not the discharger with respect to pollutants entering the MS4 from private property. The CWA imposes the obligation to obtain and comply with an NPDES permit on the person who "discharges" pollutants.[6] Where a facility is owned by one person but operated by a different person, the operator is required to obtain and comply with the NPDES permit.[7] A municipal owner/operator of an MS4 does not operate, own or control the private facilities that discharge through the MS4, and is therefore not required to obtain an NPDES permit for discharges originating on private property. The courts have also recognized the distinction between owning a conveyance and discharging a pollutant.[8] Moreover, once private property owners realize that the EPA will hold the MS4 liable for private pollution, their incentive to comply with MS4 ordinances and the CWA diminishes because they know the MS4 will be held liable for their conduct. Any liability shifted to the municipality is a forced public subsidy for the illegal acts of private parties.

Determining Effectiveness of Storm Water Management Effectiveness Plans

While the Phoenix MS4 permit does not contain numeric or narrative water quality limits, it does contain requirements to perform and report on thirty-six BMPs. These include for example, maintaining the storm drain system, paving dirt streets, prohibiting direct connections from roof drains to storm drains, collecting uncontainerized municipal solid waste, enforcing littering laws, finding and removing illicit connections to the storm drain system, and public education and outreach.

The Phoenix permit also contains requirements to monitor and report storm water quality, and to assess the effectiveness of the BMPs. Reporting on the extent to which Phoenix implements each of these BMPs is fairly straightforward. Knowing whether the extent of implementation constitutes compliance or noncompliance with the permit s not. This evaluation is subjective: how much is enough? Since most of Phoenix's BMPs have been in place for decades, determining their actual impact on storm water quality is difficult at best; there is no baseline to compare against. Although educating the public about storm water pollution is a new activity, it will be difficult if not impossible to associate with observable changes in storm water quality.

Difficult Climates

Ideally, storm water monitoring programs should focus on environmental indicators that provide a realistic assessment of the aquatic ecosystem. Such indicators are difficult to apply in an arid environment. The receiving environments in metropolitan Phoenix are dry riverbeds for most of the year. The aquatic community is ephemeral and opportunistic. There is no ability to assess water quality through changes in the structure of aquatic communities, patterns of distribution and relative proportions of sensitive and insensitive species. Biological metrics such as fish diversity indices, macro-invertebrate indices, and algal communities require a reference condition and a "wet" receiving water body.

When Discretionary Management Practices Become Required Practices

The EPA's storm water regulations require the MS4 to develop a Storm Water Management Plan (SWMP) which describes the various practices and activities the MS4 will perform to mitigate storm water pollution. The NPDES permit requires Phoenix to implement its SWMP. The question this raises is to what extent do changes to any of the elements of the SWMP constitute a permit modification requiring the EPA to publicly notice and approve the change?[9] EPA Region IX considers any changes to the permit, including additions, to be modifications requiring public notice and approval.[10] Phoenix disagrees with this interpretation, and believes that the permit sets a floor above which the City certainly can go if it wishes to do more.

Thus, the MS4 is faced with a balancing act in the development of its SWMP. In order for the document to be meaningful and provide an adequate description of the Plan's elements, some amount of comprehensiveness and detail is necessary. The challenge is to produce a document that does this while not being unnecessarily prescriptive and therefore limiting the range of operational decisions. To preserve operating flexibility, Phoenix does not consider administrative reassignments to be a program modification requiring EPA approval.

Watershed Planning

Watershed planning and source water protection initiatives represent a holistic approach to managing the aquatic resources within a single hydrologically contiguous area. Urban storm water is typically only one contributor to a receiving water. Others may include agricultural runoff, point source discharges from industrial sites and wastewater treatment plants, and atmospheric deposition. Contributions from all sources are evaluated to determine where the pollutant load is coming from and in what proportion, and control strategies are developed to reflect local environmental needs and priorities.

Not all pollution sources in a watershed are created equally. Some, like industries and treatment plants, are more amenable to end-of-pipe treatment. Others, like urban storm water, lend themselves better to BMPs and source control strategies. It would, for example, be prohibitively expensive to construct a conveyance system from Phoenix's 153 major storm drain outfalls to deliver storm water to a treatment facility that may only operate 20 days a year. MS4 operators can, however, search for illicit connections to the system and remove them.

MS4 operators would be well advised to monitor and participate in any watershed planning efforts that would affect their receiving waters. If a Total Maximum Daily Load (TMDL) effort is proposed within the watershed, the MS4 should be involved as well. The TMDL process can result in reapportioning the loading or carrying capacity of a receiving water, which can translate into revised permit conditions or limits.

Regulatory Compliance Strategies

Phoenix intends to use as many storm water environmental indicators as feasible to evaluate the success of its storm water management program. These indicators, developed by the Center for Watershed Protection, fall into six categories: 1) water quality, 2) physical/hydrologic, 3) biological, 4) social, 5) programmatic and 6) site conditions.[11] Phoenix wants it to be readily apparent that the city is fully implementing its storm water management program, and is developing an annual report that should lead the reviewer to conclude this.

To help ensure this outcome, Phoenix is assembling a draft report to present to EPA Region IX compliance staff for feedback on its construction. During this process, Phoenix intends to discuss its implementation progress in general, and to try to build a relationship with the EPA staff charged with evaluating the city's compliance. Since there are no generally accepted benchmarks with which to measure the success of a municipal program, developing an early understanding of what should be reported, and how, is critical. There is, however, an advantage to not having established benchmarks. This means that a successful SWMP is still being defined by the EPA and states, and the alert MS4 has an opportunity to define its own future.

---

[1] EPA defines a major outfall as any outfall 35" or greater serving 50 or more acres or an industrial outfall 12" or greater serving two or more acres.

[2] Claytor, R.A., Jr., (1996) An introduction to stormwater indicators: an urban runoff assessment tool. Watershed Protection Techniques, Center for Watershed Protection, Silver Spring, MD.

[3] Clean Water Act Section (402(p)(3)(B)

[4] 61 Fed. Reg. 57425, 57426 (Nov. 6, 1996)

[5] New York v. United States, 505 U.S. 144 (1992); Printz v. United States, 117 S.Ct. 2365 (1997)

[6] 33 U.S.C. § 1311(a); 40 C.F.R. § 122.21(a).

[7] 40 C.F.R. § 122.21(b).

[8] Friends of the Sakonnet v. Dutra, 738 F. Supp. 623 (D.R.I. 1990); United States v. Molitovsky Cooperage Co., 472 F. Supp. 454 (W.D. Pa. 1979); United States x. Velsicol Chemical Corp., 438 F. Supp. 945 (W.D. Tenn. 1976) same except the conveyance was a pipe carrying untreated municipal sewage); United States v. General Motors Corp., 430 F. Supp. 1151 (D. Conn. 1975) (same, storm water discharge

[9] 40 CFR 122.62(a)(1) requires material and substantial permit modifications to NPDES permits to be publicly noticed prior to EPA or a delegated state finalizing the modification. 40 CFR 124.10(b)(1) requires the public notice period to be at least 30 days.

[10] Eugene Bromley, EPA Region IX NPDES Permit Writer, personal communication, February, 10, 1998.

[11] Claytor, R. and Brown, W. (1996) Environmental Indicators to assess the effectiveness of municipal and industrial stormwater control programs. Final report. Center for Watershed Protection. U.S. EPA Office of Wastewater. Silver Spring, MD.

Storm Water Pollution Prevention: Compliance Assessment of Industrial Facilities in California

L. Donald Duke, Ph.D.[1], P.E., Member
Kathleen A. Shaver[2], M.S.

Abstract

Preliminary results evaluating compliance with storm water regulations by industrial facilities in California suggests many facilities identified in broad-based databases need not comply because activities on site do not meet conditions specified in regulations. Research compares usefulness of multiple forms of communication and site evaluation in conducting facility-specific determinations required to identify the regulated community and assess evolving patterns of compliance.

Introduction

Urban runoff is recognized as a substantial and growing source of pollutants in surface waters of the United States (Line et al. 1997). Operation of industrial facilities is one category of activity known to generate pollutants in storm water runoff, but the quantitative proportion of pollutants originating with industrial activities relative to other urban activities such as transportation, commercial facilities, and residential activities is not well understood. The industrial sector is subject to permitting requirements under the Clean Water Act and subsequent state and federal regulations to control pollutants in storm water discharges. The structure of the regulations specifies which facilities are subject to the requirements based on particular activities conducted on-site, requiring site-specific determination (Duke and Beswick, 1997). Further, the regulations require facility operators to identify themselves to regulatory agencies if they are covered. For these reasons, agencies are not readily able to identify the regulated community or evaluate the degree of compliance without relatively extensive communication efforts.

This paper presents preliminary results of an ongoing investigation at UCLA to evaluate compliance by industrial facilities with the California NPDES General Permit for Storm Water Discharges Associated with Industrial Activities, or General Industrial Permit (SWRCB 1997). The research evaluates the first stage of compliance, where facility operators are required to identify themselves by filing a Notice of Intent with the California

[1] Assistant Professor, Environmental Science & Engineering Program, UCLA School of Public Health, Los Angeles, CA 90095-1772

[2] Doctoral student, Environmental Health Sciences Department, UCLA School of Public Health, Los Angeles, CA 90095-1772

State Water Resources Control Board, or SWRCB. The primary objective is to improve understanding of the current state of compliance by facilities in the state of California with a particular focus on the Los Angeles region. The research also designs, develops, and demonstrates a range of communication and verification techniques, and assesses relative efficiency and effectiveness of a variety of approaches. The research is funded by SWRCB via a grant from U.S. EPA under Section 104(b)(3) of the Clean Water Act.

Multiple Approaches to Verify Compliance Requirements and Status

The project originated with development of a database of facilities that may need to comply with the General Industrial Permit. The database uses facility-specific information on location, business activities, and Standard Industrial Classification (or SIC) to identify "potential non-filers," or facilities that have not filed an Notice of Intent to comply but may need to comply, depending on activities conducted on their sites.

Communication with industrial facility operators to help them determine whether they need to comply with the General Industrial Permit began in November 1996 with a mailing package describing the Permit, characteristics that define a covered facility, and the duties of covered facilities. Mailings were addressed to a sample of about 10% of facilities identified in the database with SICs defined as "mandatory compliance" in the General Industrial Permit.

After mailing responses were received, researchers contacted a sample of the facilities in the Los Angeles region using three communication methods. Researchers contacted facilities failing to respond using a basic telephone questionnaire to collect data about the facility's location and whether it continued to conduct business. Facility operators responding that the permit did not apply (filing a Notice of Non-Applicability) were contacted with an in-depth questionnaire, designed to determine whether the facility conducts activities that would trigger an obligation to comply. Finally, researchers conducted simple, outside-the-fenceline site visit investigations of a number of facilities selected from the sample of facilities that had received mailings.

Estimated Compliance Requirements and Status

Table 1 summarizes results of the mailing communication. Two sources predict the approximate number of facilities that may need to comply, or "potential non-filers." The estimate falls in the wide range of 22,000 (estimated using the U.S. Census of Manufactures (U.S. Bureau of the Census, 1995)) to 32,000 (estimated using a commercial database acquired from American Business Information, or ABI). Information packets were mailed to 1,321 facilities in California, of which 404 were in the Los Angeles region.

About 7% of facility operators receiving the initial mailing recognized their duty to comply, while about 50% concluded their facility did not need to comply. These proportions were quite consistent in the Los Angeles region and in California as a whole. Between 15% and 20% of the facilities on the mailing list were not located by the postal service, and an equal proportion failed to respond to the mailing in any fashion.

Table 2 summarizes results of the basic telephone communication with a large sample of non-responding facilities in Los Angeles. Telephone contact demonstrated about 60% of those facilities to exist, though contact information from the commercial database was exactly correct for less than 10% of these. The other 40% could not be confirmed by telephone to be currently conducting business.

Table 3 shows results of the in-depth telephone communication with facilities that filed a Notice of Non-applicability. This was the only communication method that allowed independent evaluation by the researchers of the likelihood of the facilities to have an obligation to comply. (These results were dependent on facility personnel's truthfulness in responding to questions.) All facilities questioned had concluded they did not need to comply. Most appeared to be correct: 58%, or 21 facilities, apparently did not need to comply. Ten were confirmed not to be industrial firms, and 11 others appeared not to be conducting industrial activities of the types that would require General Industrial Permit compliance. Only 11% of contacted facilities were believed to be subject to the General Industrial Permit. However, the researchers did not or could not make a final determination for an additional 31% (11 of the 31 facilities). Many of these were determined to be very small firms and/or operating from the business owner's home; while these may technically be required to comply, they were presumed to be of little consequence in generating pollutants in storm water runoff, and the researchers elected not to devote further efforts to determining the facilities' compliance obligation status.

Table 4 shows facility site visits verified information in the communication database to be fully correct for a large proportion of facilities. Interestingly, database information used for the mailing was accurate two-thirds of facilities failing to respond to the mailing, but for less than half of facilities responding with a Notice of Non-Applicability. It is therefore unlikely the failure to respond can be attributed to the facility being closed or incorrectly identified.

Further site visit information is valuable in assessing effectiveness of the General Industrial Permit. Of the 31 facilities visited, about half occupied one acre or less, with 16% smaller than 1/4 acre, typically urban row storefront facilities. The latter are unlikely to generate significant runoff pollutants. A substantial proportion, on the other hand, were significantly larger.

Nearly 40% of the facilities were observed to conduct manufacturing or process activities outdoors, and nearly half maintained some form of scrap pile or waste storage area. These activities are likely to generate storm water pollutants. On the other hand, shipping and receiving areas such as loading docks, if properly maintained and operated, are less likely to generate pollutants, although these activities are defined in the General Industrial Permit as sufficient to trigger the obligation to comply. Over 80% of the facilities appeared to operate such shipping areas. Nearly 25% showed no form of industrial activity.

Policy Implications and Future Research

These results will allow future evaluation of the relative contribution of pollutants to specific receiving waters by the industrial sector in general and by specific industry categories. The database can be used to assess composition of the industrial sector within selected urban regions, the number of facilities in each category conducting industrial activities, and the typical activities exposed to storm water by facilities of those categories. Then, for those watersheds, using data available from urban stream discharge monitoring programs, future research will be able to estimate discharge of selected pollutants by industrial facilities in total; evaluate the contribution of industry relative to other sources of those pollutants in municipal storm water discharges; and identify those industrial categories contributing the greatest proportion of those facilities. These categories may then be targeted by regional pollution control strategies.

Results of this research are an initial step to address the immediate questions of degree of compliance with a critical environmental regulation, and thus the effectiveness of the storm water regulations for industry. The research develops and demonstrates methods for assessing compliance rate, and will allow evaluation of relative efficiency and effectiveness of a variety of methods which require varying degrees of time and effort. Results will lead to recommendations for an efficient protocol for use by state and regional water quality protection agencies, combining methods to identify facilities with compliance obligations and to evaluate the proportion of facilities meeting those obligations. Results will also help identify industry categories with relatively high proportions of facilities required to comply, and those with relatively low rates of meeting compliance obligations, allowing efficient targeting of resources toward key industry categories.

Conclusions

Industrial activities are identified as one source of toxics and other pollutants in storm water discharges from urban watersheds. However, the proportion of specific pollutants from industrial activities relative to other urban activities remains poorly understood. The present research is viewed as a necessary first step to quantify pollutants in particular watersheds, and particular receiving waters, from industrial activities.

References

Line, D. E., Osmond, D. L., Coffey, S. W., McLaughlin, R. A., Jennings, G. D., Gale, J. A. and Spooner, J. (1997). "Nonpoint sources." *Water Environment Research,* 1997, **69**(4), 844-860.

California State Water Resources Control Board (1997). *Water Quality Order No. 97-03-DWQ, National Pollutant Discharge Elimination System, General Permit No. CAS000001, Waste Discharge Requirements for Discharges of Storm Water Associated With Industrial Activities Excluding Construction Activities.* California State Water Resources Control Board, Sacramento.

Duke, L. D. and Beswick, P.G. (1997). "Industry compliance with storm water pollution prevention regulations." *Journal of the American Water Resources Association,* **33**(4), 825-838.

Duke, L. D. and Bauersachs, L. (1998). "Compliance with storm water pollution prevention regulations: metal finishing industry, Los Angeles, California." *Journal of the American Water Resources Association.* In press.

United States Bureau of the Census (1995). *County Business Patterns 1993: California.* Document number CBP-93-6. United States Government Printing Office, Washington, D.C.

Table 1. Mail communication, November 1996 - February 1997.

	California, statewide		Los Angeles Region	
	Number of facilities	Percent of total mailing	Number of facilities	Percent of total mailing
Estimated "potential non-filers"				
Source 1: ABI, Inc.	32,194	n/a	8,317	n/a
Source 2: U.S. Bureau of Census	22,008	n/a	5,318	n/a
Information packets mailed	1,321	100	404	100
Responses to mailing				
Filed Notice of Intent to comply	94	7	31	8
Filed Notice of Non-Applicability	718	54	201	50
Returned to sender: incorrect address	221	16	76	19
No response	228	17	78	19
Miscellaneous erroneous mailings	60	5	18	4

Table 2. Telephone questionnaires, July - August 1997: Basic questions, selected facilities failing to respond to mailing contact.

	Number of facilities	Percent of total
Facility contacted, verified to exist		
All database location information correct	17	9
Not all database location information correct	78	43
Not an industrial facility--need not comply	5	3
Company contacted, declined to verify all information	12	6
Subtotal	*112*	*61*
Facility not contacted, may not exist		
No answer; no directory assistance available	8	4
Number disconnected; no directory assistance available	18	10
Number incorrect; no directory assistance available	45	25
Subtotal	*71*	*39*
Total	**183**	**100**

Table 3. Telephone questionnaires, September - November 1997: In-depth questions, 36 selected facilities filing Notice of Non-Applicability.

	SIC Category		
	Mandatory compliance	Conditional compliance	Not industrial
May need to comply: 11%	3	1	n/a
Apparently need not comply: 58% (NNA apparently correct)	7	4	10
Determination not made: 31%			
Small facility (<4 employees)--not pursued	6	0	0
Business operated from residence--not pursued	1	0	0
No phone listing, facility out of business	3	0	0
Facility personnel declined to respond	0	1	0

Table 4. Outside-the-Fenceline site visits, September - October 1997.

A. Facility location and status of pre-existing information.

	Facilities filing Notice of Non-Applicability		Facilities failing to respond to mailing outreach	
	Number of facilities	Percent of total	Number of facilities	Percent of total
Facility name and location correct	18	45	46	67
Facility name different, location correct	5	12.5	5	7
Facility name correct, location different	2	5	1	2
Facility name undetermined	4	10	8	12
Facility closed	0	0	2	3
Facility not found	11	27.5	7	10
Total	**40**	**100**	**69**	**100**

B. Approximate facility size.*

Size	Number of facilities	Percent of total (n=31)
Smaller than 1/4 acre	5	16
1/4 to 1 acre	11	36
1 to 5 acres	7	23
Larger than 5 acres	4	13
Undetermined	4	13
Total	**31**	**100**

C. Outdoor activities visible from outside fence line.*

Activity	Number of facilities	Percent of total (n=31)
Process equipment visibly in use	12	39
Exposed to storm water	11	35
Storage of scrap, disused equipment, waste bins, etc.	15	48
Exposed to storm water	14	45
Shipping/receiving areas	27	87
Exposed to storm water	26	84

D. Facilities with only one observed industrial activity.*

Activity	Number of facilities	Percent of total (n=31)
Manufacturing activities (indoor or outdoor)	0	0
Outdoor process equipment	1	3
Shipping/receiving areas	7	23
Storage of scrap, disused equipment, waste bins, etc.	1	3
Total	**9**	**29**

* Limited to industrial facilities—omits automobile salvage yards, overrepresented in the sample.

KANAWHA RIVER

Moderator, Conrad Keyes

"Management of Kanawha Basin Conflicts"
Timothy Curran
Corps of Engineers

"Coordination of Drift Management at Bluestone Lake"
Sutton Epps, J.B. Fripp, and Coy W. Miller
Corps of Engineers

Management of Kanawha Basin Use Conflicts

Timothy W. Curran[1]

Abstract

The Kanawha River Basin contains three large Corps of Engineers flood control lakes. One of them, Summersville Lake, has the fourth largest amount of water quality storage in the Ohio River Basin. Although conflict has occurred at all three lakes, Summersville has been at the center of the most intense conflict because of competition among the various recreational and other interests for the use of this large volume of water. The Corps has seen a significant reduction in outward expressions of conflict after representatives of major interest groups became active participants in the Corps' operational information-gathering process.

Introduction

The three flood control lakes in the Kanawha Basin of West Virginia, Virginia and North Carolina are all located within West Virginia, each on a tributary of the Kanawha. Bluestone Dam on the New River was completed in 1947. Sutton Dam on the Elk River was completed in 1960. Summersville Dam on the Gauley River was completed in 1966. The lakes became operational shortly after the dams were completed. All have been involved in matters of controversy or conflict since they were built. Only Summersville Lake, however, has been involved in matters that extended far beyond the locality of the dam, and that involved so large a number of interest groups.

The Kanawha Basin watershed is composed of many hydrologic units, some of which have very different

---

[1]Hydraulic Engineer, U.S. Army Corps of Engineers, 502 Eighth Street, Huntington, WV 25701

characteristics. The mountainous areas of the upper Elk River and Gauley River watersheds have high runoff rates in the spring, and therefore low baseflow in the summer and fall. Significant sections of the watershed area in the upper reaches of the New River are flat to gently sloping, so the baseflow is correspondingly much greater during the summer months. When the relatively generous baseflow of the New River declines to very low levels in the late summer and fall of dry years, the flow of the Kanawha River is augmented by spring runoff stored at Summersville and Sutton Lakes. State and Federal agencies had agreed on the basic operating procedure for low flow augmentation before the lakes were built.

Early Conflicts

As each flood control project was put into operation, the disparity between the expectations of users and the actual results of operation created most of the initial conflicts. All three projects were involved in discussions concerning improving sport fishing, both in the lake and in the tailwaters. All three projects eventually changed the authorized winter or summer lake levels in response to recreational pressure.

At Summersville Lake, early conflicts centered around recreational issues that were unique to the project. The most pressing early issue was the loss of fish through the outlet works while the lake level was drawn down to winter level. Not only did this deplete the lake of fish, but no fish survived the passage through the project's Howell-Bunger valves. The Corps of Engineers cooperated with the West Virginia Department of Natural Resources in a research effort to minimize fish loss. The Corps subsequently raised the winter pool level, and decreased the peak release level during times of high susceptibility to fish loss.

Another Summersville conflict concerned whitewater recreation. Since whitewater was not an authorized project purpose, the Corps limited its response to boaters for special releases for whitewater rafting. Few special releases were tentatively scheduled, and those who requested a release had no assurance that the release would be provided until 24 to 48 hours before the release. This conflict, at first thought to be minor, would take on national dimensions.

Early Gauley River whitewater users were not satisfied with the amount and type of cooperation they obtained from the Corps of Engineers. They thought that more days of whitewater per season was not only possible, but that the Corps was not giving satisfactory explanations for its decision process. Lake users, who wanted the lake to remain at summer pool for as long as possible, were equally dissatisfied. Low flow augmentation releases were detrimental to lake boaters, and were a mixed blessing to whitewater interests. Private boaters (kayakers) could use low flow augmentation releases in the higher ranges. Commercial whitewater boaters could use the highest low flow augmentation releases, but only with severe restrictions as to the size of craft they could use, and with an increase in the crew to passenger ratio.

Response from the Corps of Engineers

Whitewater was becoming a popular activity, not only on the Gauley River, but in many places. The number of requests for special releases for whitewater nationwide, and especially the request of the Eastern Professional River Outfitters Association for releases from Summersville Lake, prompted the national headquarters of the Corps of Engineers in 1984 to issue guidance to its regional and district headquarters on how to evaluate requests for whitewater releases. This guidance, Policy Issue No. 216, was not intended to be a blanket permission, but rather a method to consistently evaluate these requests with adequate consideration for all relevant issues and interests. This Policy Issue also acknowledged that there may be occasions when it would be in the public interest to make project modifications that would require legislation.

The local district of the Corps of Engineers, the Huntington District, was authoring a study on modifying the operation of Summersville Lake to support whitewater recreation, and was preparing to hold public hearings on the subject when Policy Issue No. 216 was published. The study managers had been using an advisory panel composed of representatives of all interest groups affected by whitewater on the Gauley River below Summersville Lake. In the study conclusions was the recommendation that the advisory panel used in the study be continued as an aid in the management of the whitewater recreational season on the Gauley River. This study was the basis of the first planned whitewater season on the Gauley River in the fall of 1984.

Drought of 1987-88

Until 1988 the conflict experienced at Summersville Lake was local in nature and related to recreation. As the dry fall of 1987 progressed most people believed that 1988 would bring relief. When it became evident that the drought conditions of 1988 were much worse than in 1987, and that the water quality of the Kanawha River was endangered, resource agencies began regular coordination meetings.

Low flow augmentation started in the middle of June in 1988. The falling lake level put the marina out of business on the Fourth of July weekend. By September 18 the lake level was about two months ahead of the drawdown schedule, which normally starts about September 9. The volume used for low flow augmentation by that time was equivalent to about an 8 week supply of water for New York City. A normal drawdown to winter level is equivalent to about a 10 week supply.

When the Corps of Engineers reduced the amount of augmentation because of a dwindling supply in the lake, the flow in the Kanawha fell below the 7Q10 design flow used by regulatory agencies. The dissolved oxygen content of the river water fell much below that needed for normal respiration of fish and other aquatic life forms. The State of West Virginia considered changing the permit criteria for Kanawha River sewage treatment plants. This would have effectively shut down the entire chemical industry in the Kanawha Valley. Economic damages due to salaries alone would have been about one million dollars a day.

The drought experienced during 1987 and 1988 caused great financial hardship on recreational businesses. It also caused a shift in the thinking patterns of both recreational interests and resource agencies. What was initially thought to be a resource that could be exploited by one or two interest groups, was seen to be a resource that must be conserved and shared by all.

A Change in Attitude

Corps water control managers asked panel members not only their position on various issues, but the reasons why they had those positions. It became evident that most people had difficulty in voicing their true needs. Once these needs were expressed, the panel was in a position to act on the problems, not just the symptoms.

In 1989 the Corps of Engineers began to encourage the various interest groups to take a more active role interacting with other interest groups. The Corps also told the West Virginia Department of Natural Resources and the whitewater outfitters' organization that the Corps would respect any agreement that the two would make, if consistent with project purposes. Effective the fall of 1990, the WVDNR and the outfitters agreed to trade two days of tailwater trout fishing season to commercial whitewater use in exchange for an additional fish stocking at outfitter expense, and in kind services for public fishing access and emergency access. The two former antagonists had willingly cooperated to obtain mutual benefits.

The Corps had made attempts to understand the reasons behind the requests from the interest groups. Once these reasons became clear, and were discussed at the yearly whitewater advisory panel meeting, the Corps then could take steps to provide solutions in the public interest, or provide reasons why it could not respond to the request. The interest groups also gained understanding of each others' needs, and of the Corps' mission and authority, and thus became more effective in stating requests and suggesting solutions.

What had begun as an information exchange meeting had evolved into a partnership. The Corps had never considered that the outfitters' need for a definite schedule was based on business imperatives. The outfitters had not understood the way that the Corps was bound by government regulations. The state fisheries resource agency had not understood the importance of tourism dollars to the Summersville area economy. The state tourism office didn't understand how to deal with environmental issues. And the state water resource office thought that whitewater was a threat to water quality.

Results

The primary work of the advisory panel is in determining the schedule of Gauley River whitewater releases, but the feedback gained from annual meetings of the panel benefit all interest groups. Since the interest groups are also players in activities in other areas of the basin, the working relationships established through the panel are beneficial to these activities. A drought study (Punnett, 1993) used a study team that reflected the composition of the panel, and used many of the actual panel members. The drought study found that certain water conservation measures that benefited whitewater also benefited water quality and downstream industry during drought periods. These measures are now being evaluated.

The panel's cooperation among interest groups is an example that others have followed. In 1997 several government agencies followed the example of the panel and coordinated their water quality data gathering and algae research activities on the Kanawha River. Each made minor adjustments for the good of all, and every one benefited by being able to use data collected by others.

America Outdoors, a national organization of outdoor outfitters, presented their Agency Partnership Award to the Huntington District of the Corps of Engineers at their annual convention in December, 1995. To a large extent, the Corps' willingness to incorporate the feedback of the whitewater advisory panel into the operational decisionmaking process was responsible for the enhancement and conservation of outdoor recreational experiences that generated the award. Outfitters consider the Gauley River to be a model of recreational management, and it is gaining a national reputation.

Conclusions

The annual meetings of the whitewater advisory panel effectively provide a forum for information exchange, complaints, suggestions and schedule planning. The panel members must develop understanding and trust before substantive changes in business processes are possible. The operation of this panel is essential for the optimum planning and operation of the million dollar per day Gauley River whitewater season.

Upper management representation at the annual advisory panel meetings is effective in establishing the sincerity of the Corps in listening to suggestions and changing past practices. The opportunity of the interest groups to voice their requests and complaints in an open forum virtually eliminates incorrect assumptions and perceptions of bias.

References

Punnett, Richard E., and Stiles, James M. (1993). "Bringing People, Policies, and Computers to the Water (Bargaining) Table." Proc., 20th ASCE Water Resources Planning and Management Division, Seattle, WA. 495-497.

Coordination of Drift Management at Bluestone Lake

Sutton Epps[1], Jon B. Fripp[2], Coy W. Miller[3]

Abstract

The Huntington District Corps of Engineers has responsibility for operating three flood control reservoirs in the Kanawha River Basin. In response to legislation contained in the Water Resources Development Act of 1992, improvements to the current drift and debris management practices at one of these reservoirs, Bluestone Lake on the New River, has been investigated. Through close coordination with state and Federal resource agencies and local constituents, a preferred option is directed towards a modification of the dam that would allow passage of material during high flows. While the study is still ongoing, the current paper will detail the coordination effort required and a description of the proposed solution.

Project Description

Bluestone Lake is located in southern West Virginia, with the uppermost reach extending into Virginia. The dam is located on the New River approximately 3.86 kilometers (2.4 miles) upstream from Hinton, West Virginia, and 104.59 kilometers (65 miles) upstream from the mouth of the New River at Gauley Bridge, West Virginia. The drainage area controlled by the Bluestone project totals 11,823.35 square kilometers (4,565 square miles) of which 28 percent is in West Virginia, 58 percent in Virginia, and 14 percent in North Carolina. At the summer pool elevation of 429.77 meters (1410 feet), Bluestone Lake has a shoreline of 45.05 kilometers (28 miles) and a surface area of 825.59 hectares (2,040 acres). The main stem of the lake extends up New

[1]Planning Engineer, Corps of Engineers, 502 8th St., Huntington, WV 25701
[2]Hydraulic Engineer, Corps of Engineers, P.O. Box 1715, Baltimore, MD 21203
[3]Hydraulic Engineer, Corps of Engineers, 502 8th St., Huntington, WV 25701

River for 17.22 kilometers (10.7 miles). The largest tributary embayment, on the Bluestone River, is approximately 4.18 kilometers (2.6 miles) long.

Bluestone Dam is a concrete gravity structure having a height above streambed of 50.29 meters (165 feet), a top width of 4.88 meters (16 feet) and an overall length of 624.23 meters (2,048 feet). The spillway is controlled by 21 crest gates measuring 9.14 meters (30 feet) wide and 9.45 meters (31 feet) high which are used to discharge water during major floods. The outlet works consist of 16 sluices, 1.73 meters wide by 3.05 meters high (5 feet 8 inches wide by 10 feet high). Six penstocks were included in the non-overflow section of the dam to provide for future installation of hydroelectric power facilities. A gated concrete trash chute, sill elevation 452.63 meters (1485 feet), was constructed as initially designed just outside the east training wall of the stilling basin. Since the trash chute was designed for use with pool elevation 454.15 meters (1490 feet), it has not been a useable project feature.

The Bluestone Lake project was authorized for construction in 1935. The construction period began in January 1942 and continued until 1949 with a three year interruption during World War II. The authorized project purposes were flood control and hydropower. Upon completion, the project purpose of recreation and fish and wildlife conservation had been added through Congressional authorization. Bluestone Lake was the first flood control project within the Kanawha River Basin. Since that time, Sutton Lake and Summersville Lake have been developed for flood control.

As initially designed and partially constructed, hydropower was a major project purpose in terms of storage allocation. With conservation pool at elevation 454.15 meters (1490 feet), sixty percent of project storage was devoted to hydropower. However, extensive wartime electric power development resulted in a review of the Bluestone project to consider phased development, and a decision was made in 1945 to defer power development at the project and use all available storage for flood control. This decision was influenced by the tremendous increase in potential flood damages caused by the rapid industrialization and development of the Kanawha River valley and the fact that Bluestone would be the first flood control reservoir in the basin. Subsequently, the current conservation pool, elevation 429.77 meters (1410 feet), was adopted for the summer period.

Drift and Debris Problems

In 1978, the National Parks and Recreation Act (Public Law 95-625, 92 Stat. 3544) designated the 85.28-kilometer (53-mile) reach of the New River from Hinton, West Virginia (approximately 4.83 kilometers [3 miles]) downstream of the dam) to the vicinity of the U.S. 19 bridge near Fayetteville, West Virginia as the New River Gorge National River. The National Park Service (NPS) manages this National River area.

Although the existence of drift and debris was an existing condition when the National River was authorized in 1978, it gradually became clear that those who support downstream river use and development of the New River for tourism have increased the expectation that the problems should be solved. In addition, complaints and much attention have also focused on the large amounts of drift and debris (12.14 - 20.24 hectares [30 - 50 acres]) that periodically accumulate on the lake behind the dam during storm events.

Most drift and debris occurs naturally, primarily in the form of woody material, however, modern-day drift and debris also contains a man-made component. At Bluestone, man-made debris is approximately one percent, by weight, of a typical "event", but there is significant aesthetic impact when such material accumulates or is scattered about in a natural, scenic location. The debris problem is caused by human behavior, such as the act of dumping trash and household garbage on stream banks. During two of the past eleven years, there were no flood events large enough to float the dumped material downstream. During such lengthy intervals between flood events, there can be no doubt that a build-up of debris occurs until the next significant rainfall cleans out the upstream tributaries.

The movement of natural drift on waterways is normally associated with flood events as rising water over the floodplain floats decomposing woody parts of trees and shrubs and other natural organic material into the streamflow. The size of the storm event is important - larger floods collect more drift and debris. The partially decomposed condition of the woody material suggests that much of it has been passing down the waterway system for a number of years with periodic stops on streambanks and floodplains. The deposition of floating drift along the waterway is caused by the configuration of the stream channel and the character of the streambank and floodplain. In some instances, trees growing on a low depositional bar at the mouth of a tributary will provide an obstruction to floating drift. Some of the floating material also accumulates where stream curvature causes trees along the streambank or trees growing on low-lying floodplains to obstruct drift flow. Man-made debris is added to the waterways of the region as flood-flow cleans tributary streams of the household garbage placed there by local residents.

The Bluestone Lake project, as presently configured, interferes with the timely downstream passage of drift and debris on the New River. However, in this situation the material is simply delayed and then released downstream after the flood has passed. This action intensifies the downstream drift and debris problems.

Original Drift and Debris Management Plan

The original design of the Bluestone project would have provided a conservation pool elevation of 454.15 meters (1490 feet) which is 24.38 meters (80 feet) higher than the existing summer pool. A trash chute was constructed to allow drift and debris to be passed without significant accumulation behind the dam. More specifically, drift and debris would be passed, in small quantities upon arrival at the dam, through a gated trash chute with a sill elevation of 452.63 meters (1485 feet). Of more significance, the drift and debris would be passed during periods of high flow when drift normally appears in streams as part of the flood event. With this original plan, the project would not interfere with the downstream passage of drift and debris.

Current Drift and Debris Management Plan

The 1946 revision of the project operational plan substantially changed the capability for handling the drift and debris. No physical adjustment of the project was provided to maintain normal passage of drift and debris during high flow periods. With the revised plan, the material is releasable only through the low-level sluices. As flood storage occurs, the higher pool level also lifts the floating material so that it cannot pass through the low-level sluices. The floating drift and debris accumulates during flood storage and is later passed through the dam. With the pool change, it became necessary to pass this material through the dam, with manual assistance from project personnel, when the pool is at conservation elevation. This results in the debris passing as slug flow under moderate to low river flows. Debris buildups then occur in the downstream river reaches that would not have occurred under natural conditions where the majority of the debris would be transported under high river flows.

When large amounts of drift and debris are released

from the lake under these low outflow conditions, problems develop along the New River. The velocity of the water flow is less and there is greater tendency for snags and localized build-up along the stream bank. Several operational refinements have been implemented in recent years. However, the basic approach to drift and debris management is not easily modified with the existing physical arrangements at the dam.

Study Coordination

Based upon the problems identified with the current drift and debris management program at Bluestone Lake, two planning objectives were established. These objectives were: (1) minimization of the future accumulation of drift and debris on Bluestone Lake and (2) minimization of the future accumulation of drift and debris at problem areas along the downstream reach of New River through the National River.

With the initiation of the evaluation study in October 1993, coordination with agencies and organizations outside the Corps was undertaken. Coordination meetings were held with U.S. Congressman Rahall's staff, the National Park Service, the U.S. Fish and Wildlife Service, the West Virginia Division of Natural Resources, the West Virginia Division of Environmental Protection, local officials, and recreation and environmental advocacy organizations. A specific effort was made to include these agencies and organizations in discussions of drift and debris concerns and ideas for solution. All suggestions and comments during the evaluation study were carefully considered and incorporated as appropriate.

During the initial phase of the evaluation study, coordination and cooperation among participating agencies were instrumental in obtaining a better understanding of the issues. From these meetings, five key issues were surfaced. These issues concerned: (1) The release of drift and debris during high flow verses the current low-flow release; (2) the establishment of a downstream clean-up program; (3) the need for the release of some woody drift to sustain the river environment; (4) the establishment of a public awareness program designed to address deposition of debris in regional floodways; and (5) the cost sharing break down for the preferred solution.

Upon completion of the evaluation report and coordination meetings, a draft cooperative agreement was developed and signed by all participating agencies. This agreement identified the preferred project modification and management plan and the obligations of each agency for implementation of the recommended plan. The draft agreement recognizes that implementation of any or all features of the preferred plan depends upon annual budget priorities and funding actions by the Federal Administration and the State of West Virginia.

Preferred Plan

The preferred plan for drift and debris management at Bluestone Lake includes several key elements:

(1) Intake Tower. This element involves construction of a multi-level intake and a sluice through the dam. The intake tower has multiple gates arranged so that drift and debris can be passed through the dam at any lake level between elevation 433.43 meters (1422 feet) and 451.71 meters (1482 feet). The objective of this structural modification is to provide capability to pass drift and debris as it reaches the dam so that downstream passage will occur during periods of high streamflow.

(2) Implementation Actions at Bluestone Project. Operations during periods of flood storage will require an intensive effort to remove some man-made material, cut any long material for easy passage through the intake, and secure material that can not be processed. To achieve this intensive work effort, two self-propelled work barges will be purchased.

(3) Downstream Clean-up of Solid Waste. As a supplemental element to assure that all solid waste deposition along the shoreline of the National River can be addressed, provisions for downstream clean-up are included in the plan.

(4) Public Awareness Program. The primary cause of the solid waste problem in the waterways of the region is the use of roadsides and hillsides adjacent to tributary streams for disposal of solid waste. Therefore, a 3-year initial effort will be undertaken to establish a public awareness program. This effort will include seeking the participation of a broad range of agencies, organizations.

Conclusions

The results of the Bluestone Drift and Debris Study concluded that a modification of the Bluestone project through a combination of structural and operational changes, downstream clean-up and development of a public awareness program was the preferred plan. Through close coordination with resource agencies, a plan acceptable to all parties was developed and a draft agreement of participation was signed. The project is an excellent example of how resources agencies can work together.

AGRICULTURAL RUNOFF

(panel discussion related to problems of controlling agricultural runoff)

Moderator, Conrad Keyes

Thomas F. Donnelly
National Water Resources Association

Robert Wayland
Environmental Protection Agency

Mark Maslyn
American Farm Bureau Federation

J.D. McMullin
Des Moines Municipal Water District

APALACHICOLA, CHATTAHOOCHEE, AND FLINT

Moderator, Neil Grigg

"Water Resources Decision Making: A Shared Approach"
John K. Graham
Corps of Engineers, Mobile

"A History of Shared Vision Modelling in the ATC-ACF Comprehensive Study"
Richard N. Palmer
University of Washington

"The ACF Process – A Regional Perspective"*
Kathryn J. Hatcher
University of Georgia

*Text not available at time of printing.

WATER RESOURCES DECISION MAKING, A SHARED APPROACH

by

John Keith Graham[1]

ABSTRACT

Water resources conflicts have long existed in the water rich Alabama-Coosa-Tallaspoosa (ACT) and Apalachicola-Chattahoochee-Flint (ACF) River Basins. A new approach to water resources management within these river basins is being attempted. This paper describes the approach.

INTRODUCTION

In early September 1897, three steamboats loaded with cotton ran aground in the Chattahoochee River about 30 miles south of Columbus, Georgia. The *Columbus Enquirer Sun* reported on September 29, 1897, "The oldest inhabitant cannot remember a time when the Chattahoochee was as low as it is now." Columbus citizens believed the low water was caused by upstream Atlanta's construction of a water system that diverted seven million gallons a day from the Chattahoochee River. Some four million gallons was used for domestic purposes and three million for flushing her sewers. The *Columbus Enquirer Sun* exclaimed, "This thing is already becoming quite a serious matter with us and it is high time some steps were being taken to see what can be done about it."[2]

---

[1] Project Manager, ACT/ACF River Basins, Planning and Environmental Division, US Army Corps of Engineers, Mobile District, 109 St. Joseph St., Mobile, AL 36628-0001.

[2] *Perilous Journeys: A History of Steamboating on the Chattahoochee, Apalachicola, and Flint Rivers, 1828-1928*, Mueller.

More recently, a series of droughts occurred in the basins during the years 1981, 1986 and 1988. Each of these droughts significantly affected hydropower production, navigation channel reliability, municipal and industrial water supplies, and other water uses. The 1981 drought saw record low pool levels (20 feet below normal) in Lake Sidney Lanier; a major, heavily visited Corps of Engineers reservoir located north of Atlanta, Georgia. During the 1986 and 1988 droughts, numerous municipal and industrial users implemented for the first time both voluntary and mandatory water conservation measures. This sequence of droughts heightened the awareness of the finite quantity of water resources existing in the basins.

Acting upon requests by entities in Georgia to obtain additional municipal and industrial water supply storage within Corps of Engineers' reservoirs in Northwest Georgia (Lake Lanier, Lake Allatoona and Carters Lake), the Corps of Engineers, in October 1989, issued draft water supply reallocation reports proposing the reallocation of storage from hydropower generation to municipal and industrial water supply in these lakes. The draft reallocation reports contained water demand and economic analyses as well as documentation of environmental evaluations conducted under the provisions of the National Environmental Policy Act (NEPA). Shortly after public notification of the proposed reallocations, the State of Georgia, in an unrelated action, applied to the Corps of Engineers for a Section 404 b. permit to construct a water supply reservoir -- the West Georgia Regional Reservoir -- on the Tallapoosa River approximately five miles upstream of the Alabama-Georgia state line.

Serious concerns and objections were expressed by stakeholders, interest groups and the general public in meetings throughout the basins regarding the proposed actions. The prevailing concern was that Atlanta would incrementally increase water withdrawals from the basins in years to come resulting in significant adverse downstream environmental and economic impacts. Many charged that historical water resources decisions in the basins had been made piecemeal and that cumulative impacts of all potential water resources actions should be evaluated and determined before further decisions were made. The news media sensationalized the "Water War" and the conflict became a significant political issue in the Alabama Gubernatorial election of 1990 and in some Congressional races. This thing had become quite a serious matter and many felt it was high time some steps were taken to see what could be done about it.

On June 28, 1990, the State of Alabama filed litigation in the Federal District Court for Northern Alabama challenging the adequacy of the Corps of Engineers' conduct of the NEPA process regarding the proposed reallocations. Shortly after the litigation was filed, representatives of Alabama, Georgia, Florida and the Corps of Engineers began discussions seeking to resolve the issue. There was general agreement among the parties that litigation was the least desirable option for resolving the water resources conflicts. A Joint Motion to Stay Proceedings was filed on September 14, 1990, and an Order of the Court was issued on September 19, 1990, granting the Stay.

In a ceremony held on January 3, 1992, the Governors of Alabama, Florida and Georgia and the Assistant Secretary of the Army (Civil Works) jointly signed a Memorandum of Agreement (MOA) which culminated approximately 18 months of intensive negotiations. Among the provisions of the MOA were the following:

1. The parties committed to a process for cooperative management and development of regional water resources.

2. The parties agreed to participate in a comprehensive study of the water resources of the basins as equal partners and to contribute monetary and other resources to the study process.

3. A "live and let live" concept was adopted for water utilization while the study was underway. This concept included a notification procedure for proposed new or increased water withdrawals. The Corps of Engineers agreed to operate the Federal reservoirs in the two basins to maximize water resources benefits.

THE ACT/ACF COMPREHENSIVE STUDY

The purposes of the ACT/ACF Comprehensive Study are to develop applicable technical information for water resources management and to recommend a formal coordination mechanism for the long term management and use of the water resources to meet the environmental, public health and economic needs of the basins. The objectives of the Comprehensive Study are:

1. Determine the quantity of surface water and groundwater available in the basins.

2. Conduct a comprehensive assessment of present and future water demands.

3. Recommend a permanent coordination mechanism to assure shared water resources management and decision making.[3]

---

[3] *Volume I, Plan of Study, Comprehensive Study, Alabama-Coosa-Tallapoosa and Apalachicola-Chattahoochee-Flint River Basins*, Technical Coordination Group, January 1992.

The Comprehensive Study is being conducted in full partnership among the States of Alabama, Florida and Georgia and the Corps of Engineers. Representatives of the states and Corps of Engineers serve both at technical and executive levels in conducting and managing the study. Although many of the elements of the study are completed, the current schedule is to finish the remaining study elements and publish a final report by September 1998.

INTERSTATE WATER COMPACTS

The Coordination Mechanism element of the Comprehensive Study provided information on various coordination mechanisms implemented in other river basins throughout the country to resolve water resources conflicts. Primary among the mechanisms investigated were interstate water compacts which have been implemented for various river basins in the western United States and in the Delaware River Basin.

In August 1996, Alabama, Florida and Georgia agreed to develop interstate water compacts for the ACT and ACF River Basins. Each of the three states have water interests in the ACF Basin while Alabama and Georgia share the waters in the ACT Basin. Negotiations among the states and the Federal Government, represented by Department of Justice, were conducted through the remainder of 1996 and were concluded in January 1997. The proposed interstate water compacts were then introduced in each of the states' 1997 legislative sessions. The Governors of Alabama and Georgia, by satellite hookup, jointly signed the ACT and ACF Compacts in February 1997. The Governor of Florida, due to a later legislative session than Alabama and Georgia, signed the ACF Compact in April 1997.

In a joint letter dated May 14, 1997, the Governors of Alabama, Florida and Georgia submitted the compacts to the states' Congressional delegations. On June 27, 1997, the compacts were introduced into Congress. The compacts were passed by the Congress and sent to the President on November 7, 1997. The President signed the compacts into law (P.L. 105-104 and P.L. 105-105 for the ACF and ACT Basins, respectively) on November 20, 1997.

Several of the major provisions of the ACT and ACF Compacts are as follows:

1. Each river basin compact establishes a Commission consisting of State Commissioners and a Federal Commissioner. The State Commissioners, who are voting commissioners, are the governors of the respective states while the non-voting Federal Commissioner is appointed by the President of the United States.

2. Each compact specifies that a water allocation formula will be developed by the parties to the compact and that the State Commissioners shall agree

to a water allocation formula by December 31, 1998, unless this date is extended by a unanimous vote of the State Commissioners.

3. The Federal Commissioner shall, within 255 days of unanimous agreement by the State Commissioners on a water allocation formula, provide a letter of concurrence or nonconcurrence to the Commission regarding the acceptability of the water allocation formula.

4. Upon the request of the Federal Commissioner, representatives of any Federal agency may participate in any meetings of technical committees of the Commission at which the basis or terms and conditions of the allocation formula are to be discussed or negotiated.

5. Each compact establishes a dispute resolution procedure consisting of an attempt by the Commission to resolve the issue followed, if necessary, by non-binding mediation.

The compacts afford broad latitude in developing the water allocation formulas for each basin. A water allocation formula is defined by the compacts as "*...the methodology, in whatever form, by which the ... Basin Commission determines an equitable apportionment of surface waters within the ... Basin among the ... states. Such formula may be represented by a table, chart, mathematical calculation or any other expression of the Commission's apportionment of water pursuant to this compact.*"

In order to assure that the Federal Commissioner would be provided adequate information as a basis for a letter of concurrence or nonconcurrence to the Commission(s) regarding the water allocation formula within the negotiated 255-day period following unanimous approval by the State Commissioners of a water allocation formula, House Speaker Newt Gingrich, who facilitated the final negotiations on the compacts, requested that the Federal agencies develop an Interagency Management Plan (IMP). The IMP was to list the actions required to evaluate the allocation formulas, present a schedule for completion within the specified time period, and estimate the cost to perform the evaluations.

INTERAGENCY MANAGEMENT PLAN

The IMP, completed in July 1997, describes the Federal actions determined necessary by Federal law to evaluate water allocation formulas and other actions required to implement the allocation formulas. The states voiced concern that the Federal agencies would likely take a prolonged period of time following the State Commissioners' agreement on water allocation formulas to perform their evaluations and, consequently, the decision by the Federal Commissioner. The IMP presents a schedule to complete all required Federal actions within the required timeframe.

These actions include an Environmental Impact Statement and accompanying Record of Decision for each basin, preliminary flood control analysis and water supply reallocation studies for certain Federal reservoirs. Additionally, the IMP describes potential tasks that may be needed to develop a concept for long term basin monitoring and water management activities.

The Environmental Impact Statements would be prepared to address three needs: (1) describe existing system and reservoir operations; (2) evaluate the effects of potential water allocation formulas; and (3) address the needs for water supply storage reallocations within Corps of Engineers reservoirs in the upper reaches of the ACT and ACF River Basins. Since the water allocation formulas may result in changes to existing Corps of Engineers reservoir operations, the basinwide effects of proposed water allocation formulas and potential impacts to existing reservoir operations will be addressed in the NEPA documents. Each EIS and Record of Decision is scheduled to be completed in August 1999. The Federal Commissioner would be required to provide a letter of concurrence or nonconcurrence in early September 1999 if the State Commissioners agree on water allocation formulas as scheduled by December 1998.

Additional studies and evaluations to be completed within the 255-day period include preliminary flood control analyses and water supply reallocation studies. The Corps of Engineers will determine the downstream effects of utilizing existing flood control storage in certain Federal reservoirs for water supply or other purposes. Since flood control was not addressed in the Comprehensive Study, this information will be developed early in the evaluation process in order to be available as needed in developing the water allocation formulas. Water supply reallocation studies would determine the feasibility of using certain existing Corps of Engineers' reservoir storage for municipal & industrial water supply. This element would determine the economic justification and environmental acceptability of any proposed reservoir storage reallocations as well as the amount and value of existing reservoir storage space required to supply the needed water supply withdrawal.

CONCLUSION

The goal of developing mutually agreeable water allocation formulas for the equitable apportionment of water in the basins represents a significant challenge; however, the parties such an approach provides the greatest opportunity to settle the water conflicts. The states and the Corps of Engineers have been diligent in performing the Comprehensive Study in full partnership and in developing and successfully legislating the compacts for each basin. Through the ACT and ACF Compact Commissions and the ultimate development and agreement to water allocation formulas, shared water resources decision, in the strictest sense, can become a reality.

A History of Shared Vision Modeling in the ACT-ACF Comprehensive Study: A Modeler's Perspective

Richard N. Palmer[1], Member ASCE

Abstract
This paper documents the participation of the University of Washington (UW) in creating "Shared Vision Models" for evaluating water management alternatives in the Alabama-Coosa-Tallapoosa/Apalachicola-Chattahoochee-Flint River (ACT-ACF) basins. A brief introduction to the study is followed by a discussion of the history of water resources modeling. A summary of the water resources issues faced by the modeling team is presented with the strategy used in model development. The development and evolution of a series of models, denoted as the Mock Models, the Water Balance Models, the Performance Trial Models, and the Track-Two Models, are briefly described. Final comments are then provided as a summary.

Introduction
In July of 1994, the Department of Civil Engineering at the UW was requested by the Institute of Water Resources (IWR) to participate in a study of the ACT-ACF river basins. IWR and the UW had participated cooperatively previously in the National Drought Study (IWR, 1991), conducted by the US Army Corps of Engineers (Corps), developing a planning approach and computer modeling philosophy appropriate for the resolution of complex and contentious water planning conflicts. It was believed that the techniques developed in that study would prove useful in resolving the conflicts the ACT/ACF.

A Comprehensive Study, initiated in January of 1992, was conducted jointly by four study partners, US Army Corps of Engineers, and the states of Georgia, Florida, and Alabama. The purpose of the Comprehensive Study was to develop a plan for the management of all water resources in the region, to assess existing and forecasted water resources needs, and to develop an appropriate mechanism for implementing the plan. To achieve these goals, a "Basinwide Management" task was defined, with the objectives of comparing the water resources needs and availability, applying appropriate planning principles, screening and prioritizing alternatives, and building

1. Professor of Civil Engineering, University of Washington, Box 352700, Seattle, WA, 98195-2700, Email - palmer@u.washington.edu

a shared vision model to support the prioritization process. A Working Group, with two representatives from each of the Study Partners, was established to construct the models with the aid of the UW.

Use of Simulation in Water Resources

When first used in the 1950's, computer models of water resources were constructed by a relatively small number of highly trained individuals who straddled the professions of computer programming and hydro-engineering. However, by the 1980's, the environment in which water resources decisions were made began to change dramatically. The significance of environmental issues increased, and the need to increase the number and diversity of participants in the decision making process grew dramatically. These changes resulted not only in the increased importance that the public placed in environmental concerns, but also in the financing and the permitting of new projects. These changes in the institutional arrangements associated with water resources planning and management created a need for new and different approaches in the modeling of water resources. The successful incorporation of simulation models in the water resources planning process (that is, creating models that support decision making and have quantifiable impact on the decision making process) has been of particular interest.

A large number of easy to learn and apply simulation tools have been marketed that allowed the development of complex water resources models with significantly less effort than using traditional programming languages. Although the details and implementation schemes of these tools differ, most contain two important similarities which are denoted herein as an object-oriented modeling approach: 1) graphical interfaces in user-friendly environments and 2) modularity in program construction.

Shared Vision Modeling

The shift in planning and management methodology and the availability of object-oriented software created a new atmosphere in which simulation models could be developed and used. During the National Drought Study, a seven step planning procedure was defined, based upon procedures previously outlined in the "Economic and Environmental Principles and Guidelines for Water and Related Land Resources Implementation Studies" (U.S. WRC, 1983). In addition, the contributions that highly interactive, object-oriented models could make were defined and tested. The combination of a disciplined planning approach interwoven with the proper use of models was denoted as "Shared Vision Modeling." Developing a consensus of purpose in a study team was not unique to the National Drought Study; however, using computer models to focus and maintain this shared vision was.

The core of the Shared Vision Modeling approach as applied in the ACT-ACF study was to develop simulation models of the two basins that could serve many purposes. The models were designed to: 1) Catalog important data (hydrologic information, demand data, supply data, and other important information); 2) Characterize the physical features of the basin; 3) Document system operating policies; 4) Evaluate alternatives; 5) Elucidate trade-offs between operations; and 6) Expand the number of people who understood system operation. The Shared Vision Model should aid in the clear definition of planning objectives, allow users to investigate potential trade-offs (including physical, economic, and ecological) and help define suitable system management given the system constraints and objectives

The term "Shared Vision" implies a disciplined approach for incorporating managers, stakeholders, and operators into model construction. The model should contain all data necessary for accurately and appropriately representing the system under investigation. All participants should have the opportunity to review the model formally if it is to be effectively incorporated into meaningful decision making.

Issues Modeled in the ACT-ACF Study

The models developed in the Basin-Wide Process address specific planning issues that must be resolved in the study. The manner in which these issues were addressed varied over time, as will be illustrated in subsequent sections. However, throughout the modeling effort, two questions were posed at all times, "Who will use the models?" and "How will they be used?" The models were constructed by design to address management questions related to a number of issues. These management issues are described in detail elsewhere, and include: 1) Navigation Reliability; 2) Power Generation; 3) Atlanta's Water Supply Apalachicola River and Bay Water Quality; 5) Water Related Recreation; 6) Chattahoochee River Water Quality; 7) Interbasin Transfers; 8) South Georgia Irrigation; and 9) Potential Growth in Alabama's Water Demand.

Shared Vision Model Development in the ACT-ACF

The first efforts to construct a model of the ACT-ACF basins by the University of Washington came before their official involvement in the study in 1993. Rivenbark (1993) constructed a simple model of the system in a two-month effort and explored a series of operational alternatives for the basin. This was not a Shared Vision approach in that interactions with basin managers and decision-makers were limited. However, the model illustrated the feasibility of the approach to the basins, and served as a point of departure for future efforts. He developed the model in the STELLA® II (HPS, 1997) software, a highly-iterative, object-oriented software.

Following the completion of Rivenbark's model, IWR received funding for the construction of a Mock Model of the ACT/ACF. The purpose of this effort was to construct a model that would illustrate fully the potential of using the environment, and the process of beginning the Shared Vision modeling. This effort began in July of 1994 and was completed in six months. In January of 1995, this model was demonstrated at a two-day workshop to over 80 stakeholders. At this workshop, concern was raised that the simple model developed by Rivenbark had grown too large and complex for non-technical stakeholders to understand. This conflict between simplicity and detail, clarity and accuracy is one that impacts all model development, but one that is particularly important in the development of models that are used and reviewed in an open process involving a large number of stakeholders.

Based on the concerns voiced at the January of 1995 workshop, the UW embarked on an effort to create less complex models, which came to be known as Schematic Models. This work was completed by May of 1995. The Working Group rejected these efforts, believing the models were too simplistic to contribute to the process of evaluating alternatives. The Working Group suggested that the UW team work with representatives of Alabama and Georgia Power to demonstrate that the models could include hydropower operations in more detail. This next generation of models became known as the Water Balance Models. The purpose of the Water Balance Models was to simulate well-defined and realistic reservoir operations for the basins. In the end, it was concluded that the Water Balance Models should be developed to replicate, to the extent possible, published operating policies and licensing agreements. In addition, the models had to be capable of mimicking human operator nuances if this information could be supplied to the UW team. The Water Balance Models that were developed captured operating policies based upon both the Water Control Plan and licensing agreements, and applied them where appropriate based upon past operations.

Having demonstrated that the Water Balance Models could accurately move water through the basins based on past rules and experience, in January of 1996 efforts were initiated to transform the Water Balance Models into tools that would be more useful for evaluating alternatives to the Corp's Water Control Plan and hydropower licensing agreements. These models were denoted as the Performance Trial Models. The concept was to allow a user the opportunity to select a specific use or objective that was to be maximized in the basin. Operating rules were incorporated into the models to best meet these uses or objectives. For instance, a user might select a Performance Trial in which navigation targets were met. By making this selection, the model would operate to meet this objective to the extent that basinwide conditions allowed. The term Performance Trial suggested a run of the model that would calculate how well that objective could be met for conditions and operations specified. These models were developed in response to comments that the ability to investigate alternative operations in the Water Balance Models was limited.

As the study progressed, the Study Partners began to see the difficulty in defining specific alternatives before the September of 1996 study deadline and expressed interest in what was termed a Track-Two approach. In this approach, the Shared Vision Model was modified so that it could be used to help formulate as well as evaluate alternatives The Study Partners wanted this because they felt they would not be able to begin formulation of alternatives until after September 30. This approach required the Shared Vision Model to be extremely flexible. Rather than containing specific alternatives to meet study goals, the models had to help Study Partners construct alternatives by combining a number of operating features in a generalized modeling framework with suitable controls for formulating, evaluating, and refining alternatives.

The versions of the ACF and ACT models provided at the completion of the UW involvement in the Basinwide Process were designed and implemented to achieve the goals of the Track-Two Approach. These models allow the Study Partners and interested stakeholders to explore a wide range of model inputs (such as projected demands at three levels for five time periods), operational policies (influencing navigation, hydropower, meeting water demands, etc.), system configurations (variable boundaries for conservation pools, additional reservoirs, use of re-regulation storage, etc.), and system impacts (physical, economic, and environmental). The models are large by necessity, however, the user-interface to these models has been designed and organized to allow users the ability to modify the model control variables and parameters easily.

By September of 1996 two study partners, Georgia and Florida, and one stakeholder, Alabama Power, had assumed a major role in implementing modeling changes and generating alternatives to be evaluated in the model. During the final months of the UW involvement, Georgia and Florida provided valuable inputs into model changes and suggested directions in model functionality.

Conclusions

In September of 1996, the UW's involvement in the ACT-ACF study ended, model documentation was completed, and model User Guide was distributed. The three states and the Federal government agreed to enter into an interstate compact on each basin, the first ever in the Southeast. The compacts became law when the U.S. Congress passed them in December 1997. During its development, the model had been both praised and criticized, for being too complex and being too simplistic, for attempting to do more than any model could do, and for not doing enough. An accurate evaluation of the model's role in raising appropriate questions about water management and developing promising alternatives will be possible only after the deadline for the creation of stateline flows agreements under the newly created interstate compact organizations in each basin has passed.

When the invitation to apply shared vision planning was issued to IWR and UW by the partners, the partners said that their efforts to plan, rather than litigate the future of their shared resource was about to fail. If the new compacts allow the states to avoid expensive and undesirable adjudicated resolution of the conflicts, then the partners work, including but not limited to the shared vision planning, can be credited with the success.

What can be concluded now is that incorporating the models into a long and extended debate on water management proved a considerable challenge. Model development was only one part of a larger process. Each iteration of the models required first contacting those in the basins familiar with system operation, then programming their concerns into the model, then an extensive set of reviews by the Study Partners. In the estimate of the modelers, these efforts increased development time by approximately 300%; however, the efforts increased the acceptability of the model at its completion.

Acknowledgements

The author acknowledges the help of everyone who contributed to the development of the models of the ACT-ACF river basins including: William Werick of IWR; Keith Rivenbark, Lise Johannesen, David Meyer, Joe Trungale, Alan Hamlet, William Rowden, Dennis Mekkers, and Sumaya Haddadin of the Department of Civil Engineering, Janet Starnes, Ruben Arteaga, and Steve Leitman of the Northwest Florida Water Management District, Nolton Johnson and Owen McKeon of Georgia Department of Environmental Quality, Tom Littlepage and Bob Grasser of the Alabama Department of Economic and Community Affairs, Eric Nelson, Ed Burkett, and Memphis Vaughan of the Mobile District of the US Army Corps of Engineers, and Trey Glenn and Charles Stover of Alabama Power.

References

High Performance Systems, Inc. (1997). STELLA® II Tutorial and Technical Documentation . High Performance Systems, Inc., Hanover, NH.

Institute of Water Resources, The National Study of Water Management During Drought, Report of the First Year of Study, February, 1991.

Keyes, A. M. and Palmer, R.N., An assessment of shared vision model effectiveness in water resources planning, *Proceedings of the 22nd Annual National Conference, Water Resources Planning and Management Division of ASCE, Cambridge*, Massachusetts, May 1995, pp. 532-535.

Palmer, R.N., A.M. Keyes, and S.M. Fisher, Empowering stakeholders through simulation in water resources planning, *Proceedings of the 20th Annual National Conference, Water Resources Planning and Management Division of ASCE,* Seattle, Washington, April 1993,pp. 451-454.

Rivenbark, K., Reallocation of Mutil-Use Reservoirs, MSCE Thesis, University of Washington, Seattle, Washington, 1993.

U.S. Water Resources Council, 1983. Economic and Environmental Principles and Guidelines for Water and Related Land Resources Implementation Studies. Washington, D.C.

THIRD PLENARY SESSION

Moderator, William Whipple, Jr.

"Back to Basins: Nature Shows the Way"
Robert Wayland, III
Director, Office of Wetlands, Ocean and Watersheds, Environmental Protection Agency

"Legal Aspects of New Institutional Approaches to Coordination"*
Donald Elliott
Paul, Hastings, Janowski and Walker, Washington, D.C.

"Institutional Arrangements: Adjusting the Present to Future Needs"
Neil Grigg
Chairman, Department of Civil Engineering, University of Colorado

"Reorganization to Obtain Coordination"
William Whipple, Jr.
Greeley-Polhemus Group

*Text not available at time of printing.

Back to Basins: Nature Shows the Way

Robert H. Wayland, III[1]

Abstract

There is an exciting evolution underway in water quality protection and water resources management, away from single-objective, single-agency, single-facility efforts towards collaborative and comprehensive management pursued on a watershed (catchment or basin) approach. EPA has been promoting such a paradigm shift in its water programs, and assisting its state partners to do so for several years. These efforts are receiving additional emphasis under the Clean Water Action Plan commissioned by the Vice President. The environmental and policy basis for this change, how it alters some of the traditional roles of EPA, states, tribes, industry, local government, and EPA's commitment to provide technical and financial incentives to broaden and speed the change that is underway will be described.

While the '60's and '70's attempts at such approaches were ahead of their time, the widespread adoption of BAT controls for point source dischargers, the need for other means to effectively address runoff pollution, the rise of the watershed council movement at the grass roots level, efficiency considerations, and the emergence of new modeling and information management technology, have all created a more hospitable climate for these changes. In addition, the success of early models of watershed management -- the Chesapeake Bay Program and

---

[1]Office of Wetlands, Oceans & Watersheds, Environmental Protection Agency.

National Estuary Program -- provides valuable lessons that can be applied in pursuing the more pervasive application of watershed management.

The presentation will describe how EPA has "realigned" several key aspects of the programs authorized under the Clean Water Act to remove barriers to watershed management, how the Agency has supported complementary State program redirections, and key features of the Clean Water Action Plan commissioned by the Vice President on the 25th anniversary of the Clean Water Act.

The development, release, and planned upgrades for the "Surf Your Watershed" and "Index of Watershed Indicators" sites on the World Wide Web are summarized, and their role in more effectively communicating information to the general public and professionals on the condition and vulnerability of 2,111 watersheds at the USGS eight-digit Hydrologic Unit scale. Also described is the desktop modeling software, BASINs, which EPA is making available to states and tribes to allow them to more efficiently address their responsibilities under the "TMDL" provisions of the Clean Water Act.

Institutional Arrangements: Adjusting the Present to Future Needs

Neil S. Grigg, F. ASCE[1]

Abstract

The paper describes options for improving coordination between water development and regulation in a highly interrelated society. Finding workable coordination arrangements is, in the writer's opinion, the most critical water policy issue in the United States. The general options for improving coordination are presented, but specific problems will not yield to generalizations. They require specific, workable mechanisms, which may not be the same everywhere. The nation needs a more efficient approach than it has, because both our standard of living and the future of our environment require it. Promising methods for coordination are available, but the writer is skeptical than any universal method will be developed. Rather, he advocates making regulatory programs more flexible and developing appropriate forums for specific problems.

Introduction

In the arena of water policy, the most urgent need is for improvement of institutional arrangements. But which institutions? The word "institution" has different meanings, including reference to organizations, rules, relationships, behavioral patterns, and other societal customs. This paper will focus on institutions for coordination, and improving them has been identified repeatedly as an urgent need.

The Long Peak's Group (1992) suggested reforms in water policy at the beginning of the Clinton Presidency. They wrote: "... make most effective use of government and strengthen incentives for private action; integrate decisions and actions at lowest levels where problems are posed and impacts felt; integrated

[1] Professor and Head, Department of Civil Engineering, Colorado State University, Fort Collins, CO 80523.

resource management to consider demand reduction, supply enhancement, full consideration of economic and environmental costs, with full public participation; federal agency organization to promote efficiency in decision-making, consistency in administration, and public understanding of federal actions."

More recently, the Western Water Policy Advisory Review Commission (1997) wrote: "...piecemeal solutions will not do; that fundamental changes in institutional structure and in Government process will be required; that regionally-and-locally tailored solutions will be required; that the establishment of a national policy of interagency coordination which cascades down to the regional offices and field personnel will be required; and that basin-wide budgetary coordination to stimulate true integration of all Federal water activities in each locale will be required."

Finding the right balance between coordination and regulation, or as some would see it, ending the stranglehold that over-regulation has placed on water managers, is an important topic of this conference. Thus, the purpose of this paper is to describe options for how coordination could be improved. In my other paper at this conference, I described how the institutional setting which frames our water resources management process evolved. Basically, it is a story of increasing complexity in the coordination function of government in an interrelated society.

Institutional issues

Our problem is the result of increasing recognition of the rights of groups and value sets in water decision-making. In the era of water development, economic values prevailed, but in the regulatory era, environmental and other values focused on the rights of interest groups have triumphed, at least to the point of creating what some consider to be gridlock. So institutional issues that deal with arrangements for coordination would focus on finding a balance in meeting the perceived needs and claims of the different groups and value sets.

We must remember that while "to coordinate" means to harmonize, it can also imply control, and some stakeholders in water do not want to be harmonized with others. The forces that are willing to splinter to obtain their goals may be stronger than those that want to harmonize, and therein lies the fundamental problem we are trying to solve.

Making a list of the institutional issues I mentioned above, we would have: effective use of government, incentives for private action, integrating decisions and actions at lowest levels, demand reduction, supply enhancement, consideration of economy and environment, full public participation, efficiency and consistency in federal decision-making, public understanding of federal actions, regionally-and-locally tailored solutions, interagency coordination down to regional offices, basin-wide budgetary coordination, and rollback of over-regulation. These are all very general issues, and it is hard to see how progress can be made without specific proposals. However, this is at least a laundry list of what we mean by "institutional issues." Note that a number of them involve coordination issues.

Coordination theory

One missing ingredient in most discussions about coordination is what we can learn from other fields. It is my opinion that the problems of coordination are more general than our narrow view of the water field. In probing this issue, I found that a special research center has even been organized at the Sloan School of Management to study coordination. While the center is mainly devoted to information science, lessons apply more broadly. For example, the center's definition of coordination, managing dependencies between activities, applies to our discussion here (Malone and Crowston, 1994). It is this dependency structure that gives us trouble in sorting out water issues, that is, the "tragedy of the commons" is that we all depend on the same water, but we cannot agree on how to share it.

In searching for a theory of coordination in water, it sometimes seems that all we have is generalizations and philosophies, and I have been as guilty of that as anyone. As engineers and scientists we like an inductive approach, that is, we see a specific problem, and we want a general solution, and tend to philosophize; whereas politicians will tend to want a specific action fix for a specific problem so that constituents will give them the credit and reelect them. So, in searching for a mechanism to coordinate, and thereby harmonize, we are seeking to level the power base of groups so that decisions can be made for the benefit of all, a commendable goal, but one fraught with snares and traps.

Alternatives for a paradigm

So what mechanisms to coordinate have been proposed? I cannot exhaust the subject in this short paper, but I can mention a few ideas. First, let me review my own ideas about coordination.

First, I identified coordination as one of the four primary tasks of the water industry, the other three being service provision, regulation, and support (Grigg, 1996). I suggested that coordination is the weakest link in the water industry web. In Grigg (1993) I advocated a new paradigm for coordination in the water industry: "Without improved coordination ... there will be excessive conflict, wasted funds, and unnecessary damage to the environment in the US. A new paradigm for water industry coordination to be implemented within the present institutional climate is proposed with six elements: recognition of the integrated nature of the water industry; a national water management report; coordination of water management in geographic areas; national water policy studies; coordination of water data and research; and a broad array of education and training programs. ... This is a shared public-private responsibility that cannot be resolved by a single national water coordination organization."

In (Grigg, 1996), I wrote that: "Large-scale water problems require management actions by many players, but without coordinated frameworks for action, there will be gridlock, high legal expenses, and conflict-filled decision processes. However, implementing and sustaining coordinated frameworks is extremely difficult." I attempted to show in the paper that useful attributes of management frameworks should be based on inclusion, process, and control and authority. I also noted that "Neither the management frameworks nor their attributes guarantee success against negative incentives of water industry players, and success also requires water service providers and regulators to take on, in addition to their principal roles, extended water citizenship roles to help solve large-scale problems."

If I sum up the ideas above, they are that coordination is badly needed but difficult, that there are plenty of negative incentives, and that coordination is a broadly shared responsibility, not government's role alone.

William Whipple (1996), organizer of this conference, has written about coordination. In my discussion of his paper, I wrote that he had presented controversial points of view that would, in his words, "... be appropriate for comprehensive planning reviews of major river-basin systems." He recommended that the national Water Resources Council or some other equivalent coordinating agency be revived to be the arbiter of planning processes, with appropriate appeal channels available. I wrote that, in my opinion, the nation is not ready for approaches that rely on the central power of the Federal Government, because they swim against the "mistrust of government" tide that is now coming in. I suggested instead a coordinated public-private partnership approach to problem-solving. Still, a central authority is required to coordinate things; otherwise, the processes remain chaotic, as they are now.

Leonard Dworksy and David Allee (1997) have advocated a reconsideration of the concept of interagency committees. They wrote: As a result of our review of the basin reports and the special reports prepared for the Commission, we continue to be concerned about the limited consideration we believe has been given to the issue of options for State-Federal institutional arrangements, both in the regions and in a linking Washington counterpart. We understand that this is but one of the numerous issues that the Commission has to consider.... In our view it is one of the two or three overriding issues that must be considered very carefully."

Of course, watershed councils are hot in America now, for a number of reasons, but so some extent because EPA has been promoting the "watershed approach" to water quality management. In addition to watershed councils, interagency committees, and a centralized water resources council approach, other arrangements that could be reviewed would include river basin forums, shared vision modeling, effluent trading, politically-appointed, committees, compacts and partnerships, the environmental impact statement process, the recovery planning process under the Endangered Species Act, and EPA's management planning process under the National Estuary Program.

Conclusions

To conclude, I believe that finding workable coordination arrangements is the most critical water policy issue in the United States. Embedded in this goal is the goal of finding the right balance between coordination and regulation, or as some would see it, ending the stranglehold that over-regulation has placed on water managers. How to do this and still respect the rights of the many stakeholder groups in water is the key questions, given that the forces that are willing to splinter to obtain their goals may be stronger than those that want to harmonize.

We have a pretty good understanding of the general options available for improving coordination, including a revived national Water Resources Council, interagency committees, management planning processes, river basin forums, watershed councils, compacts and partnerships, coordination mechanisms in geographic areas, national water studies and reporting, politically-appointed committees, coordinated public-private partnership approaches, coordination of data and research, shared vision modeling, the environmental impact statement process, the recovery planning process, effluent trading, and water service providers and regulators taking on mandated coordination roles.

There are so many specific problems that our usual generalizations and philosophies won't help much. We need to go to specific, workable mechanisms, which may not be the same everywhere. In any case, the nation needs a more efficient approach than it has now, because both our standard of living and the future of our environment require it.

From the list of coordination approaches we have available, there are surely some promising methods that will work and help out in one case or another. I am skeptical than any overarching new method will help in most cases, however. Rather, to me, the most promising future is one of making regulatory programs more flexible and reasonable and developing the appropriate public-private, intergovernmental forums for specific problems, as they arise. This will require broad-based political leadership, expertise from agencies, the private sector, and academia, and continuing education to help citizens understand the issues and options.

References

Dworksy, Leonard and David Allee, "An Institutional Reader on Regional-State-Federal-Local Inter-Agency Coordination: An Institutional Design to Strengthen Water Resources Management in the United States," a document given to support their testimony before the Western Water Policy Review Advisory Commission, September 18, 1997, Denver, Colorado.

Grigg, N. (1993): New paradigm for coordination in water industry. American Society of Civil Engineers, Journal of Water Resources Planning and Management, Vol. 119, No. 5, 572-587, September/October.

Grigg, Neil S., A Coordinated Framework for Large Scale Water Management Actions, American Society of Civil Engineers, scheduled for 1996.

Grigg, Neil S., Water Resources Management: Principles, Regulations, and Cases, McGraw-Hill, New York, 1996a.

Long's Peak Working Group, America's Waters: A New Era of Sustainability, Natural Resources Law Center, University of Colorado, Boulder, December 1992.

Malone, Thomas W. and Kevin Crowston, The Interdisciplinary Study of Coordination, ACM Computing Surveys, 1994 (March), 26(1), 87-119

Western Water Policy Advisory Review Commission, Proposals for New Governance of Watersheds and River Basins, Draft Final Report, Denver, Colorado, 1997.

Whipple, William Jr., Integration of Water Resources Planning and Environmental Regulation, Journal of Water Resources Planning and Management, Vol. 122, No. 3, May/June, 1996, pp. 189-196. Discussion and Closure published May/Jun 1997, Vol. 123, No. 3, pp. 197-198.

Reorganization to Obtain Coordination

William Whipple, Jr., F ASCE[1]

Abstract

This conference has shown that lack of coordination is a serious national problem. We must now consider how coordination can be achieved without either delegating all power to the states or creating a Federal superagency. Legislative action will be required to authorize what may be called a "Fed-State Study" to explore the alternatives and policy problems, with a Federal advisory council to be called the "Fed-State Water Coordinating Council," to recommend the appropriate solution to the Congress. This would allow continued progress towards solving the nation's water problems, while still preserving desirable environmental conditions.

You have had abundant proof in this conference that there are many serious problems in the United States due to lack of coordination between environmental protection and development of water resources. In many river basins these problems result in prolonged stalemates and/or conflicts between agencies and interests concerned. It is not a question of which side is right in these disagreements. In many cases, both sides are right, depending upon the point of view. We must find effective ways of

---

[1]Principal, Greeley Polhemus Group, 395 Mercer Road, Princeton, NJ 08540.

preserving the environment while at the same time assuring sufficient water and electric energy to support our growing population and economy, and protect it against damaging floods and other disruptions. What we need is cooperation, but we must have procedures to facilitate it.

You have heard Neil Grigg summarize the efforts that are being made to obtain that cooperation with our present institutional arrangements. Also Don Vonnahme has emphasized the emerging state role, and the importance of interstate river basin organizations. Also the Corps of Engineers and EPA have expressed their thoughts on obtaining cooperation. Unquestionably these efforts have been successful in a number of cases. Unquestionably also in many cases current attempts have been unsuccessful, even after years of effort and the expenditure of large sums of money. In these cases, either there are states with different objectives and interests, federal agencies with established procedures which make it difficult for them to work jointly, or federal objectives differing from those of the states. Besides these differences of principle, where a number of powerful agencies are involved, there is the natural human tendency of strong individuals, each with an urge to express his or her own opinion and to see it prevail. Americans are like this.[2]

What I am now going to outline is an approach which I call a "Fed-State Study." This approach is outlined in more detail in a recent book (Whipple, 1998). The procedure would have to be authorized by Congress, and each study in an important river basin would also require specific congressional authority. Major problems would need to be solved; and where some funding would be required over and above the normal funding for planning activities of the agencies, a lead agency would be designated, with authority and funding to make the necessary arrangements. The Corps of Engineers, or in the West the Bureau of Reclamation, could readily make such arrangements. Participants in the Fed-State planning would be the federal agencies and states concerned, with other major interests brought in as appropriate.

---

[2]A characteristic not restricted to Americans.

In these studies, after preliminary meetings, computer experts from the agencies and states would be assembled to work together. This approach to using computers in multiple-agency planning situations has been developed and applied recently by researchers at the University of Washington (Palmer, *et al.*, see list of references). This technique, noted as "shared vision modeling" was developed during the National Study of Water Management During Drought and applied extensively in the ACT-ACF Basinwide Study.[3] A shared vision model is a highly interactive computer model of a resource conflict, in which managers, operators, and stakeholders are actively involved in the development as well as use of the model for decision making.[4] The model uses graphically based computer simulation environments to develop easily understood simulations of the systems under study and facilitates the testing and collaborative use of the model by all those involved in the process. The advantage of these "shared vision models," as the name implies, is that consensus in the model results can be reached more readily, since all parties participated in the development and operation of the model.

In addition to computer studies, workshops and other studies would be conducted, with each agency and state advancing alternatives best calculated to achieve its desired objectives. Finally, after discussion, the key alternatives would remain, including one or more best calculated to achieve water resource development, and others best calculated to preserve wetlands, endangered species and outdoor life for humans. At this stage, a serious effort would be necessary to determine whether there could not be found some technical improvement, such as a reregulation reservoir, which might allow adequate development of both water resources and environmental aspects. After this, each participating agency and state would be

---

[3]Alabama-Coosa-Tallapoosa and Apalachicola-Chatahooche-Flint River Basin Combined Study.

[4]Stakeholders are entities outside of government, such as power companies or water companies, which may have expertise and an interest in the study.

required to state an opinion publicly as to which alternative was preferable. (For purposes of such a study the absolute mandates of environmental statutes and/or regulations of state law would be mentioned but not considered as conclusive.)

The report with its various diverse opinion would then be sent to the executive office of the President. It would be referred to an advisory board, which might be called the Fed-State Water Coordinating Council. This council would call public hearings, and recommend to Congress which alternative should be authorized. This recommendation would then be sent to the appropriate committee of Congress for final action. Congress would authorize the plan selected, which would then be funded in the usual manner. (Obviously the organization of a congressional committee for this purpose would be a serious matter since none of the existing committees has wide enough authority.) The use of an advisory council for such purposes would be somewhat similar to the Water Resources Council, which functioned successfully for a number of years in the past.

If such a council were authorized, it would also be useful if problems arise between EPA and the states in administering the new programs of runoff control from municipalities. This program has not yet caused major difficulty, because of the caution with which EPA has administered the very great powers conferred by existing law and regulations. If pushed by lawsuits from environmental activists, the EPA may well be pushed to apply more strict controls than it has so far, in which case controversy may become acute.

There are many parts of the country where the water resources situation badly needs serious restudy. If the Congress does not move to establish the system of cooperative multiagency studies and the Fed-State Water Council as outlined, the situation will undoubtedly continue to deteriorate. A good example is the ACT-ACF Basinwide study,[5] where, after diligent and well funded efforts with

---

[5]Alabama-Coosa-Tallapoosa and Apalachicola-Chatahooche-Flint River Basin Combined Study.

our present institutional arrangements, no solution was arrived at. As long as years of normal rainfall continue, the only noticeable change will be a gradual increase in the usage of water for water supply and irrigation, and a corresponding slow decrease in low flows downstream and into the Apalachicola Bay. Nonpoint source pollution will also increase. Then one year a major drought will occur. There will be a drastic reduction in flows downstream, an inrush of saltwater into the oyster beds, widespread drying out of wetlands, and probably devastation of numerous endangered species. No extra storage or water conservation programs will be available to cope with such an emergency. Navigation will be seriously impaired. Cities will run short of water. Rivers will be more polluted. Predictably, there will be an outpouring of public indignation including environmental interests. Thereafter, in all probability, a hastily derived plan will be adapted in a rush. This projected scenario is not meant to cast aspersion upon the federal and state officials or the public in the Southeast. Past history indicates that a serious unpredictable disaster can reverse the previous conclusions of planning and public opinion in very short order, in various parts of the country. Spectacular illustrations in the Columbia and Ohio River Basins can illustrate this point. However, it is far better not to wait until a crippling emergency arises but to take informed action to select and implement the best plan in advance.

All in all, the system of Fed-State studies and the Fed-State Coordinating Council could provide a medium for resolving these extremely troublesome water resource/environmental problems without superseding the authority of any state or federal agency, and, on the other hand, without organizing a superagency with complete powers over the whole field. If we do not adopt something like the Fed-State approach, we must face the prospect of increasing difficulties in attempting to solve such problems with current institutional arrangements. This approach would allow reasonable compromises between bolstering our national economy and preserving our national environment.

Subject Index

Page number refers to the first page of paper

Author Index

Page number refers to the first page of paper